Warship 2018: Procurement of Future Surface Vessel

London, United Kingdom
11 - 12 September 2018

ISBN: 978-1-5108-8323-9

Printed from e-media with permission by:

Curran Associates, Inc.
57 Morehouse Lane
Red Hook, NY 12571

proceedings.com

Some format issues inherent in the e-media version may also appear in this print version.

Printed by Curran Associates, Inc. (2019)

The Royal Institution of Naval Architects
8-9 Northumberland Street
London, WC2N 5DA
United Kingdom

Phone: 44 207 235 4622
Fax: 44 207 259 5912

publications@rina.org.uk

Additional copies of this publication are available from:

Curran Associates, Inc.
57 Morehouse Lane
Red Hook, NY 12571 USA
Phone: 845-758-0400
Fax: 845-758-2633
Email: curran@proceedings.com
Web: www.proceedings.com

CONTENTS

Warship 2018: Procurement of Future Surface Vessel

11-12 September 2018, London, UK

PAPERS

Sponsored by:

BMT

THE ROYAL INSTITUTION OF NAVAL ARCHITECTS
8-9 Northumberland Street, London, WC2N 5DA, Telephone: +44 (0)20 7235 4622

THE CONFLICT BETWEEN SHIP DESIGN AND PROCUREMENT POLICIES

Professor David Andrews, FREng, PhD, FRINA, RCNC, (Vice President)
University College London

SUMMARY

For the naval ship designer the aim is to produce the "best" design possible. This should mean the most effective vessel that is affordable, safe and able to be produced to programme and perform effectively through its designated life. The process to ensure this can be achieved ought to follow the outline of, specifically, the concept phase. This has been detailed in a recent comprehensive paper on the sophistication of the earliest stages of ship design (Andrews, 2018). While this will not absolutely ensure the rest of the design process is also coherent and efficient, the converse of a poorly conducted concept phase is surely a recipe for failure. However, the single most difficult issue designers of such complex vessels have to deal with is the conflict between identifying a coherent design aim and the pressures of externally imposed acquisition or procurement imperatives, not least current pressures regarding perceived affordability.

The paper draws on the author's experience, particularly for four years at the end of the Cold War, when he was the Warship Project Manager for the full Royal Navy replacement programme for a substantial warfare area, namely that of amphibiosity. This was the era before Integrated Project Teams and the UK's adoption of SMART Procurement, so subsequent procurement policies are also addressed for their impact on design practice. The effect of various procurement policies on the design of major naval vessels is addressed through three specific case studies. For these the author had a significant involvement during his ship and submarine design focused career in the UK Navy Controllerate and then the Defence Procurement Agency, before enjoying a second career in teaching aspects of ship and submarine design and researching into the nature of complex ship design.

1. INTRODUCTION

> "…multi-disciplinary, multi-million dollar naval design extravagaza, where every decision must be analysed, traded off, massaged and documented to the point that the basic design issues are often lost in the process."
>
> Benford (1979)

The timescales and very significant resources, which are invested in the course of many warship designs, dwarf those invested in even the largest and most complex merchant ships; the point was tellingly made by Benford (1979) in the above quotation. This was reinforced in the paper to the 2016 Warships Conference by authors from Michigan University (Kana et al, 2016) entitled "Why is Naval Design Decision-making so Difficult?" What is also pertinent is that the sheer sophistication of modern major warships seems to necessitate extraordinary complexity in the process whereby they are designed, acquired, operated and supported; this aspect was captured by another American naval architect, Graham (1982), who commented:-

> "… today's warships are the most complex, diverse and highly integrated of any engineering system produced on a regular basis."

The usage of the term "platform" to describe a (naval) ship is a good example as to where a general procurement attitude reveals a lack of understanding the very special nature of such complex ship designs. Thus a naval combatant can provide a physical platform for complex vehicles like helicopters (such as the very significant example of the Canadian Tribal Class providing full organic support to two SeaKing ASW helicopters (Farrell et al, 1972)). However, it is wrong to use the term "Platform" to describe a naval vessel as such. It is a whole system of systems and the further split into "payload and platform" (i.e. combat system and the "rest of the ship") is a very bad design and acquisition mind-set. It implies the former is good and must be maximised at the expense of the latter. Given the so-called "platform half" of a ship provides flotation and mobility, both of essential military worth, as well as the infrastructure for the personnel (who are "fighting" the ship) and the supporting ship services (such as electrics and chilled water) without which the combat system elements could not function at all, this shows the utter nonsense of this sub-division, in not seeing the system as a whole. Thus all the interdependent functions contribute to the vessel's capability, which means there is no unnecessary "overhead" to be minimised. This typical procurement policy mind-set then leads to false justification of reducing ship capabilities in vital features contributing to sustainability, supportability, adaptability, robustness and mission flexibility with through life cost consequences beyond the individual design. An example of this thinking is the "margin-less and short life" Type 23 Frigate programme, which is discussed in Section 4.

Despite authoring a raft of publications over the years on ship design, with a particular focus on ESSD of complex vessels (see Andrews (2018) and its references), it remains hard to separate aspects related to procurement policy from general discussion of such ship design processes. Thus Gale (2003) in a general textbook on ship design and construction largely addresses procedures used by the US Navy to go through the evolving design process to build and acceptance from the shipyard. This reflects the many papers produced by Keane and colleagues from NAVSEA on the changing processes adopted over several decades in the search for "efficient procurement" of US naval vessels (Keane et al, 2009a and 2016). Thus policies like Total Package Procurement, Design to Cost, Competition Advocacy, and Acquisition Reform (Tibbitts & Keane, 1999) were adopted and then succeeded in turn, with significant effects on the designs produced under each initiative. One representation of the way in which the constraints and pressures on a given design programme can change through the design's development was indicated by the "Push-Pull" diagram (see Figure 1).

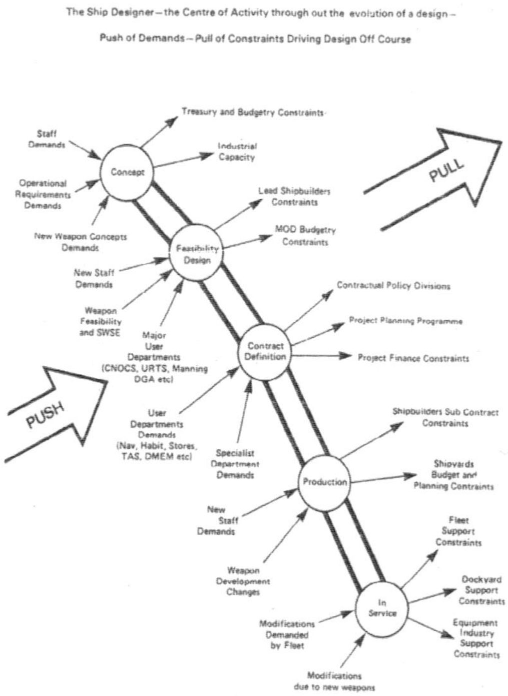

Figure 1: The Phases of ship design showing the Push of Demands and the Pull of Constraints (Andrews, 1986)

One of the significant criticisms of the use of the design spiral as the description of the ship design process, beyond the fact that ship design just isn't like that (except at the micro-level of design iteration to a weight and space balance – see discussion in Andrews (2018)), is that it appears to be a closed process. Given the comments in this paper on the effect of procurement policies on actual ship designs, this is clearly wrong. An early attempt by the author to address this was a three dimensional representation of the spiral as a "cork-screw" (to reflect the progressive and iterative nature of design) which then enabled the various external constraints to be emphasised and categorised – see Figure 2. This figure shows three broad types of constraints on the ship design process, which apply regardless as to whether one adopts this representation or any other description of the ship design process (Andrews et al, 2009):

a) Constraints directly placed on the design by an explicit statement in the requirements or later in the build specification;
b) Constraints directly limiting the design team's scope in executing the design;
c) Wider constraints, sometimes explicit but more often implied, that originate in the environment in which the design team have to operate.

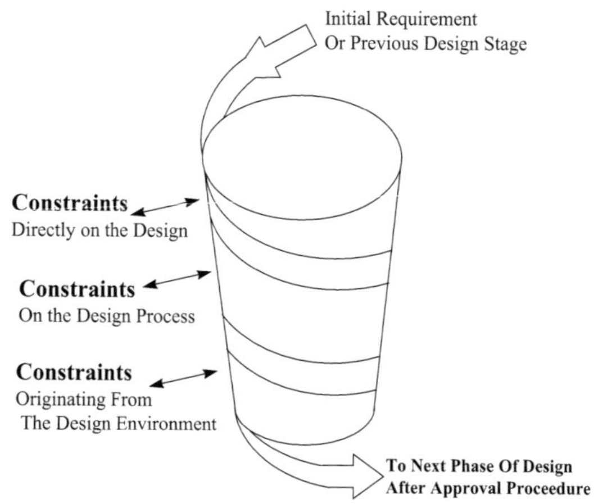

Figure 2: A three dimensional representation of the "design spiral" with three types of external constraints (Andrews, 1981)

Table 1 gives examples of each of the constraint categories. This listing from Andrews (1981) is, surprisingly, still largely valid, in general. Perhaps two significant additions to the third category can be seen to have arisen in subsequent decades, namely:

i) the adoption of commercial practice in naval ship engineering and acceptance;
j) the increasing emphasis on meeting government/international initiatives of an ecological nature.

Table 1: Three categories of Design Constraints on a Complex Ship Design (Andrews, 1981)

Direct Constraints on the Design	Constraints on the Design Process	Constraints Originating from The Design Environment
Minimise Building Time	Structure of the Design Organisation	Physical and Natural environment. Political climate
Consider foreign sales potential	Relationship of the designer with customer	Economic Climate
Reduce manpower on the ship	Attitude of the design organisation to the latest design techniques	The exact manner in which money is funded
Reduce specialised manpower on the ship	Past design type ship data available	The need to comply with new laws (eg health and safety during build)
Minimise the maintenance load required at the ship	Countries of origin of designer or design methods	The political necessity to support ailing shipyards
Simplify production process in the shipyard	The need or ability to buy-in talent to the design team	The strategic and political necessity to spread work round the shipyards
Fit up-to-date equipment which is being concurrently developed with the ship	Specialisation and training of design team	The decision to reduce direct government research
Minimise time in refit	State of the art in the various fields	Collaboration with NATO allies on equipment
Minimise time in port	Computer facilities directly on tap and their limitations	
Comply with international rules existing or likely to come into force	Quality of general engineering data directly available	
Minimise training load to operate ship	Research facilities directly under the designers' control	
	The idosyncracies, prejudices, rivalries, personalities of the design team	

The fact that such constraints can have a dominating impact on a design through direct requirements, the design team's ability and the political (in the widest sense) environment in which the design process takes place, leads on to a quite different representation of the ship design process to that of the design spiral, even in its most sophisticated representations (see Andrews et al, 2009 for some 26 iconic representations of the ship design process). Thus Figure 3 shows the whole ship design process to not be just a series of sequential tasks to be performed but rather as a sequence of major decisions that a ship designer or design organisation, but not a CAD toolset, has to make in order to proceed (see Appendix A to Andrews (2018) for an explanation of each step). Most of these decisions are taken in the concept phase, which further emphasises its importance. In fact the decisions made in the concept phase take up the bulk of this diagram, such that the subsequent phases of the design process (on this strategic basis) can be summarised as being ever more detailed design iterations, through the repetition of the last three steps of Figure 3. It is worth highlighting that this is a top level representation and that, while designer decisions ("Selection") have to be made before their related design tasks can be executed in the process, sometimes the sequence of the steps may be different when specific aspects drive a particular design.

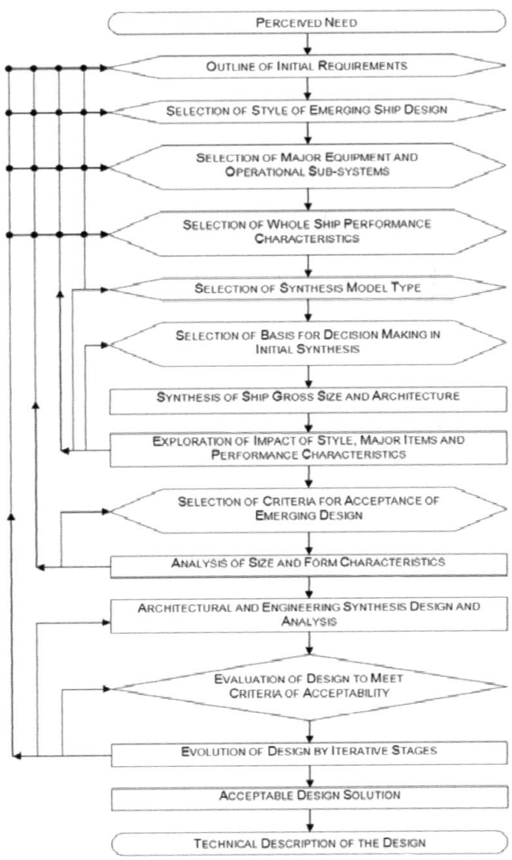

Figure 3: A representation of the overall ship design process emphasising key decisions (Andrews 2018)

Two earlier considerations of the impact of procurement policy on naval ship design are next considered. The first was produced from the author's experience as a warship project manager in the late 1980s and the second slightly later when some of the post-Cold War defence procurement "reforms" were underway. Following this retrospective look, two "shibboleths" that still affect naval ship acquisition are addressed, namely, the myth of "gold plating" and the belief in adopting commercial standards (in order to reduce (initial) ownership costs without significantly degrading military capability). A lesson on the nature of naval vessel design, which draws on the thoughts of one of the most successful naval designers and project managers of all time, is then summarised. Three examples are analysed (from the author's career in the RCNC) showing where procurement policy edicts and high level management decision making compromised the design and acquisition of those Royal Navy programmes and so provide case studies underlying the assertion of this paper. Finally some suggestions as to how the warship project manager of the future might cope in minimising procurement policy conflicts with the design intent, while recognising both the complexity that that two above quotations encapsulate means this can only ever be "a damage limitation exercise" at best.

2. PREVIOUS ANALYSES OF THE IMPACT OF PROCUREMENT POLICIES

2.1 THE ROLE OF THE WARSHIP PROJECT MANAGER

The author presented a paper in the early 1990s, which was based on the author's four years as a warship project manager (WPM) in the UK MoD Controllerate at the end of the previous decade and thus reflecting procurement concerns at the end of the Cold War (Andrews 1993). Thus the pre-occupations were prior to the changes in NATO defence budgets seeking the "Peace Dividend. The paper was entitled "The Management of Warship Design", 'though as Professor Rydill pointed out, in one of fifteen contributions to the discussion recorded in the RINA Transactions, the paper mainly addressed procedures rather than processes for this very fraught and demanding activity. Rydill considered "A procedure is a way of organising one's business and a process is a series of activities undertaken to achieve an end result", and while different in character "constitute the warp and weft of the whole operation" (of procuring a warship programme). In the author's response I saw the paper's aim as being to show where procedures affect the process – and thus, for the present paper's purpose, how the external pressures impact on the design team. This can explain why at the end a coherent ship design process is often not discernable and the resultant design and physical realisation often compromised by procedures over which the WPM has little control. Even worse are instances where senior management (up to the political levels) rip up, ignore and even subvert mandated procedures, often for over-simplistic instances of expediency.

The most relevant sections of the 1993 paper to the current theme on the effect of procurement policies are those on "the tasks of the WPM" and "the characteristics of warship project management". The former lists five over-riding concerns that were seen to pre-occupy every WPM:

- Survival of the project;
- Achieving the necessary approvals to proceed;
- Producing and maintaining the contract with the shipbuilder (and the build specification upon which the contract depended)
- Preserving the programme, both schedule and approved expenditure;
- Sustaining the project team.

Surprisingly, some senior discussers queried the first of the above, but I maintained that this was the foremost of my concerns. This was in part to protect the team below me so they could get on with the real (design) job and partly because, unlike the management above me, I did not have to juggle other competing projects – as the most senior officer devoted solely to my project(s).

The second section on characteristics made the point, which was reinforced in my subsequent positions as Head of Concept Design and then Surface Ships Project Director, that every project has different programme imperatives and also specific design drivers (see Andrews, 2018). However, sub-structuring the process to control both the design and the wider project issues (such as the procurement strategy) was also subject to much discussion, reflecting the organisational arrangements of the day. The many external checks on the project (driven largely by the necessity to control public expenditure) and the internal checks, within the wider procurement organisation, were also significant. It was noticeable, over the author's 35 year career in the RCNC within the UK MoD, how much the characteristic of the top-level management changed. Initially, there were senior staff technically involved who drew on immense domain knowledge. This was provided largely by members of the RCNC for warship through life design management. Whereas, now from my observations from outside government, there would seem to be a reliance on "domain inexperienced" professional project managers with most of the technical effort being contracted out.

The section concluded with an outline of three comparative ship procurement approaches:

- The US Naval ship acquisition approach: (This well-resourced organisation (NAVSEA) has itself undergone considerable reductions in capability, but some of these appear to be being restored (Winter, 2007, Keane et al., 2009b).);

- The French warship acquisition approach: (This seems to have changed least in the last few decades due to its significant role in defence sales which makes it less beholden to the "naval staff" - see discussion in Section 3 and Andrews (2013).);
- The specific Australian acquisition of the COLLINS Class submarines. (Much of the story on the eventual inadequate outturn on this programme has now been published (Yule & Woolner, 2008), although back in 1993 it seemed to be a well-funded and resourced prospective acquisition.)

That similar procurement environment pressures have arisen beyond the UK warship procurement domain, probably reflects the post-Cold war era, which might be about to change geo-politically. However, there also seems to be a growing recognition that the special nature of the naval procurement environment and the associated shipbuilding industry are different and challenge the generic defence nostrums that have existed for nearly half a century (Winter, (2007), Parker (2016)).

2.2 "CURRENT" PROCUREMENT ISSUES

A second paper was produced shortly after the 1993 paper and its discussion were published, for the following year's INEC (Andrews, 1994) and entitled "Current Issues in Warship Procurement" and better reflected post-Cold War defence procurement concerns. It addressed the "new defence nexus" under five "C", namely, Competition, Collaboration, (prime) Contractorship, (TL) Cost and CALS (i.e. Computer Aided Acquisition and Logistics Support). The latter item had been adopted from a US DoD initiative to control paperwork and now seems quaint in the light of the 21st Century of the globalised Internet and the Big Data environment. Perhaps the salutary message here is that we cannot predict the wider environment in which defence and naval vessel acquisition will exist and we ought thus to adopt genuinely smart and resilient organisations and practices. But not the so-called SMART Procurement (MoD 1999), which it wasn't (Andrews, 2003 & 2011). A far better approach would have focused on the real objective: "the best warships we can provide within the nation's allocated resources – and to cultivate the right people to achieve that consistently"?

In considering the remaining "C"s, they now seem more dated than those concerns of the WPM identified from the late 1980s. Competition was said to be the means of achieving VFM but over three decades on we now have not just a monopoly for at least submarine procurement but have adopted a partnership approach for the QEC ships in the Carrier Alliance. Both policies reflect a diminished number of vessel programmes having to be stretched to keep a national acquisition capability, with the issue of getting the "best" vessels through life the bottom line. Collaboration with other navies in Europe was seen as attractive before the demise of the NFR90 programme (Schaffer & Kloehn, 1991) and the experience of the CNGF (van Griethuysen, & Juliot 1996). The former was obviously a step too far since having the US Navy in the programme was unbalancing. Having been one sixth of the Anglo-French steering group setting up CNGF in the early 1990s, the author saw, in combining with the French ship design organisation (DCN), the virtue of a common ship design early on (particularly to control the aspirations of the combat system development process). However, it was always likely to founder once the quite different ship acquisition practices and work share demands of (eventual) three nations started to dominate design decisions. So the final outcome, fraught though it was, was better than the alternative in whole ship terms.

Prime Contractorship also seems based on a historic optimism. It originated with Peter Levine ("a poacher turned gamekeeper") as Chief of Defence Procurement believing that government could focus on capability needs and then throw the requirements over the (contractual) wall – this is Requirements Engineering with all its flaws once more (Andrews 2003, 2011). The US Navy has recognised (even where it still has an industry able to mount competitions) that such hands-off of the design of concurrent ships and weapons is not good value for government (Winter, 2007). Through Life Cost remains a desire consistent with good basic engineering principles but severe problems of credibility.

3. THE MYTH of "GOLD PLATING" AND THE BELIEF IN COMMERCIAL STANDARDS

It is somewhat unfortunate that very often the conclusion of the ship operator fraternity on a new ship design is a quick judgement on the worth of that new design. Thus when the successor to the very successful LEANDER Class Frigates, the Type 22 Frigates (BROADSWORD Class) came into R.N. service they were said to be branded by the Controller of the day as "painted ladies" (a term that now sounds more than a little anachronistic), in part reflecting that fact that the ships lacked a medium calibre gun. However the design was very much a Cold War ASW escort and actually the design restored decent levels of Growth (i.e. unplanned observed growth) and Board (future use) Margins both necessary from recent experience. These margins enabled that design to be readily stretched twice (Batch II and III) to accommodate new sensors and then a medium calibre gun and the C3 facilities to be a screen command ship. So that Controller was wrong, as indeed was the First Sea Lord who required the replacement of the similarly adequately designed Type 82 Class (only HMS BRISTOL was built following the axing of CVA-01) by the "smallest ship able to take (the AAD) Seadart system to sea" – the Type 42 Destroyers. The latter then exhibited all the predictable results of not adopting adequate through life adaptability, in the mistaken belief this was a means of avoiding "gold plating".

The Type 42 Class could thus be seen as the direct result of procurement pressures, resulting in an ill-thought through "efficiency saving". That the then senior ship designer (Sir Alfred Sims as Director General Ships) agreed to this policy rather than the historic response, which had been to insist that such obsession with initial cost minimisation was bad practice, probably reflected the shock of the loss of the carrier replacement programme. This "smallest ship" mind-set has unfortunately too often been the procurement policy, so the Type 22 stands out as a sensibly generous design.

The second substantial policy pressure that has occurred originated both from senior naval staff and other procurement policy authorities who had a say in the procurement of naval vessels, but without much understanding of their design principles and the nature of shipbuilding. This was a belief that commercial standards, which is itself another simplistic and ill-defined generalisation, could be readily adopted for naval vessels without any noticeable effect on the fighting (and sustainable) efficiency of warships – but would (it is always asserted) enable cheaper ships to be acquired. This is explored further through two actual ship programmes in Section 5. However it is worth saying that one of the consequences of a very minimal investment in naval ship research – certainly in comparison with aerospace vehicle design and defence wide combat systems research and development – is that such assertions frequently become policy for the next major ship programme. Too often this is adopted without any proper investigation to determine that it both reduces total programme cost (including in-service) and still provides an acceptable capability to justify the investment and commitment.

It is usually accepted that, cheaper to fabricate, merchant ship structures will be heavier than sophisticated naval hull structures, however this is rarely checked out for the overall impact on a combatant design. In such vessels the structure is significantly the largest element of weight and thus impacts on ship size in an iterative "knock-on" effect, that results in overall ship size growth (and cost growth) before a balanced design is restored. Given also that the heavier but simpler merchant-ship structure is likely to give reduced levels of ship survivability to contact or standoff underwater explosions, due to the cheaper but less integrated structural connections (Bradbeer, 2013), such a choice ought to be carefully weighed up. However all too often the cost spiral and capability degradation is not quantified before a policy decision is taken, in the hope of overall initial procurement cost saving.

This is not to say that all lessons that naval ship procurers can learn from other practice are without merit or applicability to warship acquisition. Several decades ago the North Sea oil and gas industry became aware that its profit margins were likely to be severely reduced if it continued to employ similar commercial practices to those adopted by naval ship acquisition under the Competition/Prime Contractorship practice outlined in Section 2.2. This had meant that the risk past from the customer (oil field owner in the former case and government in the latter) to the Prime was cascaded down the sub-contractor/equipment vendor hierarchy and therefore added considerable cost to the oilrig or warship. There was also no incentive for sub-contractors to suggest better/cheaper/quicker ways of assembly and installation of their "part of ship". So the oil industry introduced CRINE (Cost Reduction In the New Era) where several leading sub-contractors, the prime and the customer shared risk and savings. This led to some 25% reductions in North Sea oil installation costs. It took the UK MoD far too long to break with the "Competition" mind-set with several single source or artificial competitions (over "indicative" but not contractually binding prices) before, in the case of the QEC carrier programme, a Carrier Alliance with risk and profit sharing amongst several major contractors and the UK MoD was concluded as the best possible arrangement. Whether this was introduced too late to make a significant design difference (as had occurred in the CRINE case) must probably await the outturn on the procurement of these the largest vessels ever procured for the Royal Navy.

4. A LESSON ON THE NATURE OF NAVAL VESSEL DESIGN

Sir Rowland Baker was the most eminent naval constructor post-WWII, as argued by his biographer David Brown (1995), who was also an eminent historian on ship design, with his Centennial history of Naval Construction (Brown, 1983). Rather than either Baker's illustrious wartime activities on Mountbatten's Combined Operations staff or his final Technical Directorships of the DREADNOUGHT and then the Polaris Submarine Programme in the 1960s, what I want to do for the purposes of this paper's argument is to focus on a paper written by Baker. This was written when he was seconded, as a Constructor Commodore, to the Royal Canadian Navy to lead the project for the RCN's first indigenous ship design, that of the St LAURENT Class Frigates. The paper was entitled "How to build a ship" (Baker 1956a) and though non-technical and for general naval consumption, wonderfully captures Baker's philosophy, which reached final fruition in his very successful project directing of those two key Royal Navy nuclear submarine projects.

Baker starts by stating:

" 'So the chicken comes before the egg and the ship comes before the staff requirement' The naval staff, who must decide what the ship must be capable of doing, can only state a requirement in known terms and this means the staff requirements for a new ship must refer explicitly or implicitly to some existing ship… the shipbuilder or designer depends on the staff to say the sort of ship they want, the staff whether they realise it or not, depend on the shipbuilder and designer to indicate what sort of ship is possible."

Thus Baker is consistent with the view of the author in that a Requirements Elucidation approach is key to the Concept Phase for complex vessels (Andrews 2018). He then goes on to list six issues seen to be necessary to finally produce a new ship:

1. You must have an industry to build a ship.
2. You must have a designer, which Baker interestingly identifies:-
   i) Having the knowledge, which is to "know what he himself can do";
   ii) To be an artist, which Baker sees as not just aesthetics but almost having "feel", since: "if it doesn't look right, it cannot be right (regardless of meeting the staff requirement)";
   iii) Achieve a "suitable" arrangement of the compartments (he says more on this as is summarised below).
3. Someone must require a ship. This Baker sees as the naval staff (for the ships he was procuring), who must:
   i) Understand how to fight (It was salutary for the UK Ministry of Defence to re-learn some obvious lessons from its war to recapture the Falkland Islands in 1982 (Meek, 2003));
   ii) Have a sound knowledge of ships, sea and weapons, which has become harder with fewer ships in the fleet;
   iii) Know what both the crew and materials can do. (Probably the second is the harder for sailors to acquire?)
   iv) Have judgement, which Baker considers to be: the power of seeking advice; deciding on its quality; and maintaining decisions once made (All these issues are addressed in Section 6.);
   v) Thus "a ship when completed cannot be a total success, she can well be a total failure", which Baker considers to be the chain like consequence of one significant failure in something as complex and interlinked as a ship. He lists failure to be:-
      - The staff requirement can be wrong (see Preston's assertion "There are more bad staff requirements than bad ship designs" (Preston, 2002));
      - The design can fail to meet the staff requirement, which sounds straight forward but Honnor's adage "the best ship design is that which just fails to meet its requirement" reveals the subtlety of the issue (Honnor, 2006);
      - Construction can fail to meet the design requirement, which contractually means failing to meet the specification (which can never be perfectly watertight);
      - All too common minor failures. (This can in part be down to having no prototypes, yet still having expectations of airplane/car production line delivery, rather than ship building being more like that of the construction industry's practice with inevitable "snagging lists". The obvious solution to (often) poor ergonomics is more money for detailed design and improving the quality of outfitting, which (sadly) has not traditionally been part of the highly competitive maritime culture.)
4. Approving the arrangement provokes two further Baker-isms: "This is probably where the real skill of the designer comes in (of course he has to be a fully qualified naval architect)." and "Once everyone is allowed to explore the (n!) "alternatives", the process of design could go on forever. That is NOT the way to build a ship."

Baker's solution to the immense difficulty of preventing senior (naval) personnel from insisting that their personal design views on ship arrangements being incorporated into a new ship design was spelt out in his 1955 SNAME paper on the St LAURENT Class (Baker, 1956b). Figure 4 shows how his "Stylised" layout restricts allocation of single functions to specific areas of the layout. This was specifically adopted to ensure that "interested parties" beyond the designer, were constrained from "interfering" in the design and was devised by Baker as a design management ploy. It reveals his insight into whole ship design, where the naval architect is both the design leader and the right discipline to lead the overall project's acquisition.

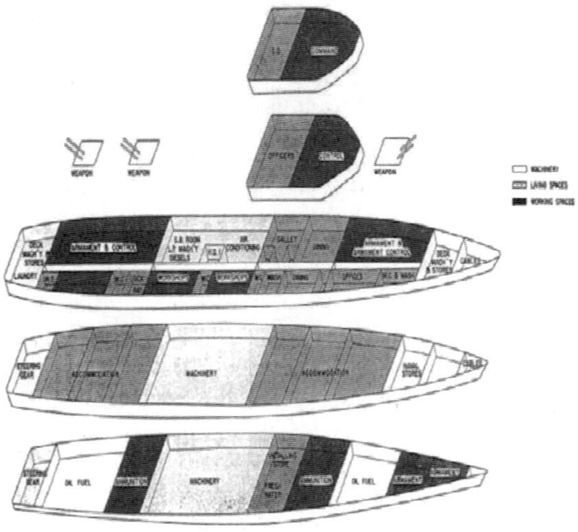

Figure 4: The "Stylised" Layout of Baker's St LAURENT Class Frigates (Baker 1956b)

Two further overviews were made by Baker, in his article, one on the traditional design control of the whole design as it went into the Detailed Design phase and the second on the need to judiciously order materials for build assembly. In discussing both aspects he very much reflected ship construction practice of sixty years ago, with detailed design approval and close involvement in the build by the navy/government acquisition agency.

One comment there is very pertinent to this paper's argument "If the hull is poor, the whole thing fails." This again puts the naval architect in the prime professional position in the ship design process, 'though it could be argued against this that greater safety concerns nowadays for the hull engineering/naval architecture aspects mean the naval architect can't now lead on the overall project too? Baker's and my response to this would be that only the naval architect can be the overall designer and splitting the wider project responsibilities, so well integrated into Baker's very successful ship and submarine programmes, from the grasp of the person responsible for the vessel's design issues too often courts disaster. In fact, Baker said as much attention needed to be paid to the plan as the design. By "plan" he meant the procurement strategy and the project programming, which show clearly that the two can't be separated and are interdependent if final success of a naval project is to be achieved.

5. SOME KEY CASE STUDIES

5.1 TYPE 23 FRIGATES

This frigate class evolved from the 105m "towed array tug" concept, which then grew in steps (112m, 118m) to 123m long general-purpose frigate post-Falklands War (and after the official Concept Phase) (Bryson, 1984). It was the first flared R.N. hull form (for radar cross section minimisation reasons – see Figure 5) and pioneered a combined diesel-electric and gas turbine (CODLAG) propulsion fit (for ultra-quiet acoustic signature to operate the towed array). Both these features were incorporated from the Concept Design studies and were the two most fundamental ship design decisions, retained from the concept studies, despite the very significant growth in size, post-Concept. Where the procurement policy nostrums of the day could be said to have had a devastating effect on the design was in the political mandate for the class to be to be short life and "margin less", when everyone in the concept team "knew", despite the Navy Minister's edict, this would not be held. Many ships in the class will now be in R.N. service for at least 28 years, rather than the mandated 18 years ship life, and the through life cost of this is obvious. This shows the impact of an ill thought through policy which was sustained as the operational requirement grew from the 'towed array tug" to being the next fleet escort and world-wide deployed combatant. So in subsequent service that "short life" policy was discarded to avoid starting the next frigate programme for over two decades.

The Type 23, for which the author was the project leader for the concept phase, was also a lesson in not designing a naval vessel to too precise a set of requirements. The operational requirement was part of a Cold War strategy of "blocking" the GRIUK gap to prevent Soviet submarines getting into the Atlantic and attacking US reinforcement to Europe. Thus the towed array role was very precise and OA was used to "optimise" array towing speed, sprint speed and patrol endurance. This made the design choices much easier than for a general-purpose escort or world-wide combatant. However such precision was negated when the Cold War ended and the class was deployed to the Gulf and elsewhere as a general combatant. Luckily its 4.5in gun had been fitted late in the design due to the Falklands Campaign. (There had been no gun requirement in the GRIUK Gap.) So this is another lesson to be learnt from an overly specific operational specification, which had ignored the inherent adaptability of naval vessels for the unexpected – but adaptability is only achievable if sufficient margins are built in for the unexpected future.

A final lesson from this programme, which reveals the design impact of decisions made on one class affecting future designs, concerns the arbitrary imposition of a reduced number of billets on the ship ("too keep the size (i.e. cost, i.e. initial cost) down"). It was arbitrary because no studies were done to justify the reduction and when the first ship got to sea it was obvious there were not enough men (in those days) to fight the ship. Some further bunks were squeezed in (thus failing the agreed standards) but even so Training and Advancement billets were lost. And the consequence was that the next class of combatants (the Type 45 Destroyers some 10 years later) had to have twice the number of these necessary fleet billets to compensate for the earlier decision. This was neither included in the cost implications of the "cheap/short life" Type 23s nor costed into the Type 45 size growth to avoid repeating the error in the future. That such errors can be minimised by a better approach to concept design, which can question such arbitrary decisions "from on high", has been outlined more than once in the author's call for a proper concept dialogue through Requirements Elucidation (Andrews, 2011, 2018).

Figure 5: Type 23 Frigate

5.2 HMS OCEAN

Commercial standards were mandated for this helicopter carrier (Figure 6) without this being assessed through any concept and feasibility studies. This arose from the adoption by senior naval personnel, not ship designers, of a false costing based on a belief (since it had yet to be delivered) that the simpler conversion of a merchant ship to a training role (see Section 5.3) could be readily extended to a major amphibious warfare vessel. The Project Manager (the author from 1986-1990) fought the naval staff over adopting such non-naval standards for an essentially high value unit (given its "cargo" of hundreds of troops plus associated equipment, including 12 Commando helicopters).

After an abortive acquisition attempt with a capped purchase price, the PM managed to raise the purchase budget but not sufficiently enough to cover the incorporation of quite limited naval standards. The eventual purchase was subsequently criticised by the Ministry of Defence's chief marine engineer as this commercial practice substantially increased the engine support requirement for the fleet (due to this one ship's unique engine fit). Needless to say, these support costs were not shown in original cost based decision, given this was obsessed with direct initial procurement cost rather than the "true ownership cost" of the design solution. Interestingly, many have argued that HMS OCEAN has been "good value for money", as it appeared to provide a substantial amphibious lift capability to a more commercial price level. However that capability has only been exercised in "peacetime roles". The ship has not been used in operations of full naval warfare (rather than usefully in peace keeping), so the jury must be out as to whether this commercially engineered vessel constitutes a "cheap" solution, which is capable of discharging a major naval capability "for real".

One of the procurement ploys to try to keep the design (and build) of all but the military upper works as commercial as possible was for the prime Contractor (BAE Systems Barrow) to have the hull and machinery built/installed in its Govern Shipyard, which had only built merchant vessels. This and a "hands off" policy by the MoD led to one clear incident that showed that building ships for the government can not be done without risk, if "hands-off". When the ship was launched the starboard forward poppet stuck and flew off. Luckily the port one followed, otherwise the ship could have toppled sideways on the slip before entering the Clyde. As a result the stem of the ship hit the concrete slip before fully afloat. The next day I observed the launch video and it was obvious that this shipyard was not used to launching naval hull forms with fine forward lines requiring tall staves at the two forward poppets (This is basic naval architecture – see Newton's classic textbook (Newton, 1956)). These staves must be cribbed together to prevent the tripping that occurred in this case but hadn't been done (as this is not usually necessary for shorter poppet staves typical for fuller merchant ship forms). Now, given the importance of this being checked beforehand by the project and its local representative (the Naval Construction Overseer in each naval shipbuilding yard at that time – since cut by staff reductions), why had this not occurred? The argument was used that the Project's contract was with the Prime Contractor (BAE Systems Barrow yard), so the launch was just a sub-contractor and Barrow's responsibility. The fact that the Crown had paid (by the usual milestone payments) for most of the ship at launch shows the falseness of such "procurement nostrums" applied to the unique nature of ship design and build. (That the Project Manager was a marine engineer and thus not aware necessarily of the dangers of dynamic ship launching was more a fault of a system that believed good project management of these unique products did not require the level of domain knowledge inherent in an experienced naval constructor.) This instance also shows why the Constraints in Table 1 are so important in their potential effects on ship design and acquisition.

Figure 6: Amphibious Helicopter Carrier (LPH) HMS OCEAN

5.3 RFA ARGUS

This ship was procured by the major conversion of a Ro-Ro containership to helicopter training ship (Figure 7). The procurement strategy could have been said to have been flawed from the start, as it was seen as a "quick win" to acquire a modern merchant ship "easily" converted to an auxiliary training role and normal procedures could be short circuited. Add-back money (underspend) had to be used quickly so the tendering process was rushed. This was too short (6 weeks + 2 weeks Assessment) and the cheaper of two proposals (by Harland & Wolff) accepted. H&W were keen to re-enter naval work, as they wanted to win the much bigger new build AOR, so this could be seen "for learning and a loss leader". H&W also lacked weapon system installation experience but this was seen as low risk as "limited weapons fit for ATS" (but it did include a new aircraft control C3 fit). H&W were contracted to convert CONTENDER BEZANT (selected by H&W) in accordance with MoD's quickly drawn up Statement of Technical Requirements (SOTR) in the ITT. The Contract proposed that MoD supplied the weapons fit (thus H&W would "just" be the installer i.e. SWSE), however in endorsing the Staff Requirement the Fleet Procurement Committee changed the contract to Whole Ship Procurement (WSP) "as a safe experiment". This meant H&W were now the prime Weapons Authority (SWSA) through a Post-Contract Weapons Amendment, "spelt out" in a Cardinal Point Specification. In reality H&W contracted to Racal that responsibility, acting as postman and were "Hands Off", so a questionable "experiment" by MoD policy makers! Finally H&W produced an inadequate Cardinal Date Programme (revised 4 times) despite MoD extending Contract Acceptance Date (CAD) from 28 February '86 to 19 December '86 "to sugar the WSP pill".

MoD Project Management (due to high level policy) was also seriously in error. It was assumed that WSP meant the MoD Project Team could be instructed to "just monitor", despite the fact that H&W consistently failed to provide Contract Delivery Plan, a Quality Plan, Progress Reports, etc.. No Naval Overseer was provided on site, however Lloyds Register was intended to be employed (for the merchant ship aspects but not the new naval aviation fit), and subsequently LR refused to work with H&W being only prepared to "assist MoD". The WPM was given no staff for this project, as the Controller said this "will prevent interference" and the main task of the WPM's team was the CVS programme, so the ATS was seen as "Low Priority". H&W sent compartment drawings to MoD "for comment" and the (CVS) project then provided considerable guidance on arrangements and detailed design (this was felt to be necessary because of the hastily produced SOTR) and H&W's drawing offices lack the confidence to challenge any of MoD "corrections". Finally DGST(N) (the RFA operating "customer") made no comments on drawings but during outfitting inspections identified many "design defects".

The Production Phase was no better as it became clear that H&W had not adequately surveyed ship at Tender (despite taking on that responsibility) so that subsequently a lot of rectification was found to be required (surprisingly so for a three year old ship but shows why it should have been fully inspected before purchase). There were 130 Alterations (excluding the WSP amendment) largely due to the inadequate SOTR, 45 Force Majure Claims (only 5 were accepted by MoD) with 45 also in the subsequent Claim. Confusion occurred over Tests, Trials & Inspections, in part due to mixed commercial/naval standards for RFAs (i.e. Naval: aviation, C3, RN accommodation; LR/DGST(N): commercial structure and outfit; RFA accommodation to civilian standards). H&W planned Tests, Trials & Inspections from June '86 to March '87 were unrealistic (some spaces had eight repeat Inspections!). Ship was finally accepted 3rd March '88 (i.e. 24 months late and 14 months after last agreed CAD revision. H&W withdrew from the Part IV Trials and the Guarantee Phase. One can only conclude that **WSP Failed**, yet MoD persisted with it as a procurement strategy.

So one can summarise from RFA ARGUS' conversion that this was a classic case of "how not to procure a naval vessel":

- ADDBACK: led to the timescale being rushed;
- An inadequate Specification was produced;
- A major conversion is never that easy;
- Having a mixture of Naval and Commercial/RFA standards was an added complication;
- Post-contract introduction of a new procurement strategy (WSP) is always foolish;
- Totally under-resourcing the MoD Project ("to prevent interference") reveals a strange view by the Controller of his own staff and VFM;
- Minimal Overseeing staff is asking for trouble;
- If a shipyard is inexperienced with MoD work then it needs close MoD project and naval overseeing supervision.

Having taken on this project in its intended acceptance phase the author, as Project Manager, had subsequently to defend to the parliamentary Defence Committee (the HCDC) the payment of huge cost overruns on a Fixed Price contract (H.M.S.O., 1989). This case proved naval ship acquisition is a lot more than just engineering design and even conversion to a support role, such as this training vessel, can be demanding. The resultant procurement failure was largely due to several unwise acquisition edicts imposed on the project team, early on in the project and resulted from a procurement policy decision, which could only be described as incoherent in the rush to get what looked like a "quick win". An important negative lesson is how crucial can be the choice of vessel and procurement proposal, in this case to substantially modify an already built merchant-ship to naval operational capability. As with the LPH procurement, decisions which ignored the advice of experienced naval ship designers and acquisition experts, are more than likely to lead to poor technical solutions, costing more than budgeted and coming into service late.

Figure 7: Aviation Training Ship (ATS) RFA ARGUS

6. HOW THE PROJECT MANAGER MIGHT COPE WITH MINIMISING PROCUREMENT POLICY CONFLICTS WITH DESIGN INTENT

6.1 BETTER COSTING

Almost all the procurement policies have underlying them the imperative to keep (in UK cases) UPC to a "minimum". Thus the cost of a new ship programme becomes excessively focused on this (frankly) artificial measure of expenditure or better "value for money". The fact is that UPC is a construct. It is not the actual total money paid out by the project to the prime contractor over the several years of monthly stage payments, particularly for the First of Class (assuming a class like that discussed in Section 5.1 rather than the single ships in Sections 5.2 and 5.3), where there will be lead items and design and individual Project Orders for development items. This is excluding any substantial sensor and weapon sub-system developments, which arise alongside the combat management system development, possibly including a substantial Shore Test Facility, which may not (see Type 23 Class) be included in the UPC (no STF but development cost in ARGUS's UPC).

Furthermore, most major decisions on affordability are made in ESSD when costing is essentially using historic data of dubious relevance to the new programme and even at the point of contract award for the new build can only be indicative, with no contractual commitment by the shipbuilder to meet it. This does not seem to be appreciated by the senior staff (and their political masters), as without full-scale (and aircraft like "flying") prototypes any cost must be. In fact until the shipbuilder (and it is the shipbuilder whether they are the Prime or not) has sufficient detailed design definition to go out to the myriad of sub-contractors and equipment vendors for "best and finals", the material costs (typically at least 50% of ship cost) are not contractually confirmable. This very intense negotiation with typically hundreds of sub-contractors (e.g. amounting to 426 in the case of the Australian COLLINS Class submarines (Yule & Woolner, 2008)), usually takes a year to conclude and can impact on the ship design, due to obsolete equipment or changing vendors for a "better deal". This leaves the construction, outfitting/setting to work and T,T&Is to be added to give the shipbuilding cost element of a contractually confirmable procurement cost.

What even this UPC, when finally assessed, does not actually directly relate to is the "value" of the vessel – it is only the fabric and part of the cost of producing it. Fuel, stores, ordinance and the considerable cost of providing the crew are omitted from UPC. Hence, as any good engineer knows, any sensible measure of value/cost should include in service or operating costs, but for a service vessel with a very un-predictable service life this is highly speculative. Despite the best endeavours of the MoD in insisting "equal weight will be given to initial (procurement) and through life costs" potential prime contractors for naval programmes "know' that the cheapest "bid" to deliver, based on UPC, will likely win the contract (thus HMS OCEAN and RFA ARGUS). Until MoD can better devise value based cost approaches that also recognise the "nature of the beast" we will continue to have over-runs, as typified by so many recent programmes. But this did not occur with Baker's projects – why? See the last section below.

6.2 PROBLEM OF SIZE AND DENSITY OF COMPLEXITY

The second issue that needs to be better addressed by the ship design and acquisition fraternity is that of getting over to the hierarchy above them, in the MoD and government, a recognition of the fact that like it or not (and unfortunately there is something about bureaucracies and governments that don't like being faced with, to them, what are uncomfortable facts) "ships are different". This is due to their shear size as vehicles and that naval ships, in particular (as the quote from Captain Graham USN says), are complex and densely so.

Some years ago, in an excellent paper on the "family of warships" that Vosper Thornycroft had successful sold worldwide, Usher and Dorey (1982) pointed out that if such ships were designed a bit more generously some of the issues of design, build (and even support through life) could be mitigated. Keane, as the Technical Director at NAVSEA for over a decade, in his discussion to Andrews (2018) highlights "adequately sizing a naval ship" since the opposite (in concept) "increases ship outfit density and adds unnecessary complexities to the design and construction of the ship". A clear demonstration of this was the Japanese "copy" of the

USN DDG51 Class Destroyers, the Konguo Class. These were re-designed to be longer (by some 6%) in order to make them easier and cheaper to build. So, why is there an obsession with "keeping ships compact" and, in particular, short? This unfortunately comes back to the poor basis of costing future ships.

The part of the build, which is reasonably (and solely) within the shipbuilders' domain, is that of steelwork (even though as a proportion of UPC it is very small, typically 12%). As well as steel, outfit costs (at a material level) will go up due to greater distribution of ship distributed systems and coverings. However, ease of installation and through life accessibility far out weigh this, as occurred in the DDG51/Konguo example and, for a naval auxiliary, the Avondale redesign of the USN LSD 41 Class later ships to optimise installation work. It can also be seen as symptomatic of a belief going back to "gold plating" referred to earlier, which reveals a lack of trust of the ship designers to not "over-design". The fact that more comfortable "first-rate" designs are invariably good value for money over parallel "second-rate" solutions (see several examples in Brown & Andrews (1980) and the argument summarised in Section 3.1 of Andrews (2017)) is too often ignored under short term budget pressures and the senior management seemingly unable to listen to sound design based advice.

6.3 THE DRIVE FOR SALES

With diminishing national naval shipbuilding programmes the still widely held belief, that one's own warships should be built indigenously, means that just building the nation's fleet results in the overheads to maintain that capability are spread over fewer hulls. So those overheads then become significant in the procurement costs for each vessel. Thus the question arises as to whether such (second tier?) naval powers ought to bias their ship designs to a wider market, rather than just the needs of the "home navy"?

It could be said from this side of the English Channel that such a "sales-driven" approach to naval procurement has been a strong and successful feature of French naval procurement. However, I think it has only been part of a wider "strategic" approach to French defence procurement, which seems to be in stark contrast to that in the UK. Thus SMART Procurement (MoD 1998) insisted that project managers (IPT Leaders) think just of their immediate projects rather than what this meant for future programmes (and even through life sustainability of current programmes) – with results of the sort highlighted in the previous section's examples. Thus when the author visited the French carrier CHARLES de GAULLE building in Brest naval dockyard and quizzed the project manager over the lack of pre-outfitting, the response was "Yes, we will build the ship slowly and more expensively this way – but there is no new ship needing to be built sooner and employing more workers (for a quicker cheaper build) would then affect the region's unemployment budget." This is thinking not just strategically for the naval programme but nationally. The same argument was used to fit two (submarine) nuclear plants on the same ship – costly for that design (and questionable project VFM) but very sensible in maintaining a critical national capability. (An interesting thought (due to a former RCNC colleague) is that "if all the material costs were indigenous then (due to tax revenues) naval ship programmes would be zero cost to the Crown". Such a view would question the national benefit of, say, the MARS Tanker programme being largely spent in foreign exchange in Korea?)

So is the logic behind the Type 31e Frigate programme, argued strongly in Sir John Parker's report to the UK government (Parker, 2017), a sensible and strategic way to deal with the above overheads or will the Royal Navy find such ships "second-rate" and dispose of them early in their lives? That is what happened to the Type 14 and Type 81 Frigates (but, notably, not the Type 23s despite being "designed for short life"). This seems potentially another example where procurement policies, so often adopted without sufficient regard to defence needs and strategic VFM, could yet again end in further loss of naval capability?

Talk of adopting a European defence acquisition policy, despite NFR90 and GNGF/Horizon experiences, could also lead to reduced capability designs, with which naval ship designers may well have to do their best. It is worth recalling that even with the national strategic French ship procurement practice, resulting in nationally highly regarded and influential ship designers, they are also sometimes prey to political direction. In the early 1990s at the start of CNGF discussions with the French naval staff and the acquisition organisation (DCN), which at first we did not expect to progress, we were told: "This (collaboration) will happen because "Monsieur le President" has told us it must." Such is the underlying "political" nature of naval acquisition, whose complexity is well captured by Benford's statement.

6.4 SHIP DESIGNERS TAKE CONTROL

The title of this final sub-section makes it clear that the conflict this paper addresses can only be adequately dealt with if the ship designer (particularly from the start of the concept work) takes control. We have been through a series of "acquisition reforms" in both the UK and in the much larger US Navy in recent decades that have contributed to ill thought out projects and accelerated the decline in fleet size. Most of them (with the notable exception of Winter's (2007) call to "take control back" by NAVSEA) have deliberately diminished the role and authority of the in-house designers of naval vessels. One would think the cause and effect would be obvious but the author (having experienced this process) has been surprised at how little this has been appreciated. This is despite him stating it directly to the Royal Navy through its main organ for naval thinking – the Naval Review's Centennial publication (Andrews, 2013).

Winter's solution, when Secretary of the Navy, was entitled "Getting Shipbuilding Right" and can be seen as the US Navy's reaction to the loss of design control, due to too much of a "hands off" approach to its warship acquisition. Three of the six steps he lists to restore control, are Items:

1. "re-assert its control of the entire shipbuilding acquisition process"; (Winter goes on to say this covers the "selection of key trade-offs – performance, crew size, logistics support, cost and schedule".)
2. "the lead systems integrator should be the Navy (which in UK terms means the NAVSEA equivalent, the DE&S ship project) – not the contractor"; and
4. "the navy needs independent cost estimates for trade-offs/decisions that we make".

All three are seen to be relevant to dealing with the conflict identified in this paper. Furthermore, the US Navy has followed Winter's call by setting in chain a basis for re-establishing NAVSEA's design control. The best summary of how this is being done is in Keane's paper "Ready to Design a Naval Ship? - Prove It!" (Keane et al, 2009), which presents three tables on the skill set and career experiences required by future NAVSEA Ship Design Managers.

So in conclusion:

1. There is nothing wrong with the strategic use of the naval acquisition programme for wider national purposes and therefore why not buy out the strategic overhead and build new warships on a marginal costs basis;
2. But the navy also needs good ship designs, which are founded on a Requirements Elucidation approach and a comprehensive Style exploration in ESSD (Andrews 2018);
3. While new combatants need the latest relevant weapons and sensors, the ship designs themselves also have to be "first rate", which means "affording" adaptability;
4. Innovation in ship design should not be suppressed as part of "minimising risk". Naval ship design is risky but naval programmes should adopt a management approach (such as Systems Architecture (Andrews 2015)) to minimise risk rather than adopting "cheap and nasty" options, which are poor VFM.

7. ACKNOWLEDGEMENTS

It should be clear that the above view of the conflict between the management of warship acquisition and ship design draws on the author's career in the RCNC from 1965 to 2000 and his continued involvement, at UCL since 2000, leading research in design and acquisition of complex vessels, yet remains his own view. However, ship design is a supreme example of group endeavour, so I both acknowledge those colleagues (particularly fellow members of the RCNC) in the UK MoD and in the naval industry, from whom I learnt so much and, latterly in my former career, those many colleagues in the several projects, where I was fortunate to be the project lead, albeit through "interesting times" (with thanks to the Chinese aphorism).

8. REFERENCES

ANDREWS, D. J. (1981), *Creative Ship Design*, Trans RINA Vol. 123, 1981.

ANDREWS, D.J. (1986), *An Integrated Approach to Ship Synthesis*, RINA Transactions, Vol. 128, 1986

ANDREWS, D. J. (1993), *Management of Warship Design*, Trans. RINA Vol. 135, 1993.

ANDREWS, D. (1994), *Current Issues in Warship Procurement*, Int. Naval Eng. Conf., Plymouth, IMarE, June 1994.

ANDREWS, D, J. (2003), *Marine Design Requirement Elucidation rather than Requirement Engineering*, IMDC 03, Athens, May 2003.

ANDREWS, D. J. (2011), *Marine Requirements Elucidation and the Nature of Preliminary Ship Design,* IJME Vol. 153 Part A1 2011.

ANDREWS, D. (2013), *The View from Bath – a Naval Constructor's Perspective*, Article in book *From Dreadnought to Daring – Centenary of the Naval Review*, Edited by Captain Peter Hore, Seaforth Pubs, Barnsley, UK, Jan 2013.

ANDREWS D. (2015), *Systems Architecture is Systems Practice in Early Stage Ship Design*, IMDC 2015, Tokyo University, May 2015.

ANDREWS, D. J. (2017), *Choosing the Style of a New Design – The Key Ship Design Decision*, IJME, Vol. 159, Part A4, 2017.

ANDREWS, D. J. (2018), *The Sophistication of Early Stage Design for Complex Vessels*, Special Edition of IJME, Vol. 160, 2018.

ANDREWS, D.J. et al (Ed) (2009), *State of the Art Report on Design Methodology*, in IMDC 2009 Proceedings Edited by Erikstad, S. O., Trondheim, June 2009.

BAKER, R. (1956a), *How to build a ship*, Royal Canadian Navy Publication, 1956.

BAKER, R. (1956b), *Habitability in Ships of the Royal Canadian Navy*, Trans. SNAME, 1956.

BENFORD, H. (1979), *The Changing Scene in Marine Design Techniques*, Symposium on Advances in Marine Technology, Norwegian Institute of Technology, Trondheim, June 1979.

BRADBEER, N. (2013), *Shock Response Implications of Lower-Cost Warship Structural Styles*, PhD, UCL, 2013.

BROWN, D. K. (1983), *A Century of Naval Construction*, Conway Maritime Press, London, 1983.

BROWN, D. K. (1995), *Sir Rowland Baker*, WARSHIP, 1995.

BROWN, D K & ANDREWS, D J (1980), *The Design of Cheap Warships*, Proc. of International Naval Technology Expo 80, Rotterdam, June 1980. (Reprinted in Journal of Naval Science April 1981)

BRYSON, L. S. (1984), *The Procurement of a Warship*, Trans RINA, Vol. 127, 1984.

FARRELL, K. at al. (1972), *The DDH-280 Design*, SNAME, Canadian Chapter, 1972.

GALE, P. A. (2003), *The Ship Design Process*, Chapter 5 of LAMB, T. (Ed.), *Ship Design and Construction*, SNAME, Newark, NJ., 2003.

GRAHAM, C. (1982) quoted in GATES, P. J & RUSLING, S.C. (1982), *The Impact of Weapons Electronics on Surface Warship Design*, Trans RINA, 1982.

van GRIETHUYSEN, W. J. & JULIOT, P. (1996), *PROJECT HORIZON Design management in a Multinational Environment*, IMarE INEC-96, Den Helder, Neth., April 1996 (Reprint NEJ Vol. 36.)

H.M.S.O. (1989), *Supplementary Estimates Vote 1, Class II: Payments to Harland & Wolff PLC*, House of Commons Defence Committee, Minutes of Evidence, Wed 29th Nov 1989.

HONNOR, A.F. (2006) quoted in ANDREWS, D.J. et al (Ed) (2009), *State of the Art Report on Design Methodology*, in IMDC 2009 Proceedings Edited by Erikstad, S. O., Trondheim, June 2009.

KANA, A. et al, (2016), *Why is naval design decision-making so difficult?* RINA Warships 2016, Bath, UK, June 2016.

KEANE, R. et al (2009a), *"Ready to Design a Naval Ship? – Prove It!"*, SNAME, JSP, April 2009.

KEANE, R. et al., (2009b), *A Human Capital Strategy for Ship Design Acquisition Workforce Improvement: The US Navy's Center for Innovation in* Ship *Design*, USNEJ, No.4, 2009.

KEANE, R.G. et al, (2016), *Reducing Detail Design and Construction Work Content by Cost-Effective Decisions in Early Stage Naval Ship Design*, JSP&D, Vol. 32, No. 2, 2016.

MEEK, M. (2003), *There Go the Ships*, The Memoir Club, County Durham, 2003.

MOD (1999), *A Guide to SMART Procurement*, Ministry of Defence Smart Procurement Implementation Team, Edition 1 –April 1999.

NEWTON, R. N. (1970), *Practical construction of warships*, Longman, London, 1970.

PARKER, J. (2016), *An Independent Report to Inform the UK National Shipbuilding Strategy*, London, 3rd Nov. 2016.

PRESTON, A. (2002), *The World's Worst Warships*, Conway Maritime Press, London, 2002.

SCHAFFER, R. L. & KLOEHN, H. G. (1991), *Design of the NFR-90*, USNEJ, Vol. 103, 1991.

TIBBITTS, B. & KEANE, R. (1999), *Acquisition Reform: It All Starts with Design,* USNEJ, Jan 1999.

USHER, P. & DOREY, A. (1982), *A Family of Warships*, Trans RINA, Vol. 124.

WINTER, D. (2007), *Getting Shipbuilding Right*, US Naval Institute Proceedings, June 2007.

YULE, P. & WOOLNER, D. (2008), *The Collins Class Submarine Story*, Cambridge U P, 2008.

9. AUTHORS BIOGRAPHY

David Andrews was given a new Chair in Engineering Design at UCL in 2000 when he retired from the UK MoD as a Director Surface Ships. He led the design integration of aircraft and weapons fits on the INVINCIBLE Class Carriers, was senior naval architect in the Forward Design Group before promotion to Warship Project Manager for the Amphibious Replacement Programme and a significant portion of the in-service fleet. He was subsequently Head of Preliminary Design in Future Projects responsible for initial design and procurement studies, including those on the Future Surface Combatant, CVF and the WAVE Class tankers. In 1993-98 he was appointed MoD Professor of Naval Architecture at UCL where he directed the MSc in Naval Architecture and led UCL's innovative research into Trimaran ship design. In 2001 he set up a new Design Research Centre in the Department of Mechanical Engineering, which is focused on computer aided ship design (CASD) and the design of complex systems. He is a Fellow of RINA, a Vice President and Member of Council. In 2000 was elected a Fellow of the Royal Academy of Engineering and is currently the International Chair for the tri-annual series of the International Marine Design Conferences.

How much is too much? The cost of commonality

O Short, Naval Design Partnering, UK
S Lee, Naval Design Partnering, UK
T Coe, Naval Design Partnering, UK

SUMMARY

Combining requirements to create common adaptable designs is widely understood to provide key cost advantages through a vessel's life. Combining requirements can provide advantages, but there is a limit to how many requirements can be combined whilst still producing cost savings. This study uses a common launch as an example to understand the main cost drivers and how these are affected with a common design. Using the main cost drivers, HiSCAT, an NDP in-house rapid design tool, is used to understand how different requirement combinations affect the overall cost at pre-concept phase. Using three sets of requirements the main cost drivers including consumables, crewing and procurement are calculated. Results are perhaps counter intuitive as they show that the most cost effective solution is not to combine all three requirements but produce two variants. This provides an initial method for conducting fast assessment at pre-concept stage to guide requirements combination and provide optimal designs.

NOMENCLATURE

The following abbreviations and acronyms are used throughout this paper;

Acronym	*Definition*
CSS	Commercially Supported Shipping
GFE	Government Furnished Equipment
HiSCAT	High Speed Craft Assessment tool
ITEAP	Integrated Test and Evaluation Plan
MoD	Ministry of Defence
NDP	Naval Design Partnering Team
NRC	Non Returning Costs
PQQ	Pre-Qualification Questionnaire
RFI	Request for information
ROM	Rough Order Of Magnitude
R&P	Resistance and Prolusion
UPC	Unit Procurement Cost

1. INTRODUCTION

It's widely understood that combining requirements to create common adaptable vessels provides advantages in many key areas including spares, operability, availability and design costs. This is especially relevant where operational fleets have developed over a long period of time resulting in an assortment of designs with varying manufactures and equipment. A prime example of this is within the Ministry of Defence (MoD) where commercially supported shipping (CSS) boats currently operate and maintain over 60 different vessels from a variety of manufacturers.

To reduce diversity, fleet rationalisation is taking place to establish a common fleet. In 2016 the MoD procured a set of work boats ranging from 11m to 18m in four size variants covering the role of 11 in service classes [2]. These designs are modular, allowing various roles to be undertaken by the same vessel.

However, there is likely to be an upper limit to the number of requirements that can be combined whilst still reducing costs, as it's not expected that all 60 classes can be incorporated into one single design. Using an example case study from the Naval Design Partnering Team (NDP), a common launch is used to answer the question 'how much is too much?'.

This question is investigated by initially looking at the through-life cost drivers for a common launch and quantifying these costs where possible. We then incorporate cost drivers into rapid design tools, looking at all possible requirement combinations, to understand the most cost effective solution.

2. BACKGROUND

The basis for this paper is a study conducted by the NDP on a common launch [3]. The common launch looked to combine requirements from six different in-service vessels currently conducting four duties into a single common design. The roles that the craft is expected to undertake are;

- Patrol
- Escort
- Dive Support
- Utility

Which are undertaken by the following in service vessels;

- Island Class
- P2000
- 14m Launch
- 15m Patrol Launch
- 16m Scimitar Class
- Condor (Arun 60)

Prior to this study, requirements were captured through duty holder engagement to define a base set of requirements for each operational role. All requirements were collated into a single data set and from this a concept design was developed. The concept ensured all duty holder requirements could be met with a single common craft. Outputs were provided in the form of a general arrangement, weight spreadsheet, resistance and propulsion calculations and rough order of magnitude costs.

Feedback provided on the final concept stated it was too expensive and complex for the majority of the operational roles. This was mitigated by providing a flexible arrangement allowing mission critical kit to be added/removed when required increasing capability whilst still offering benefits to maintenance, training and spares. Even with this flexible arrangement the concept was deemed unsuitable.

Therefore, the initial aim of the task presented here was to understand whether, even with an over engineered vessel, are the cost benefits of a common design enough to outweigh a simpler craft, or whether requirements needed to be broken down further into multiple designs.

3. COST DRIVERS

To understand the potential savings that can be made by a common craft, the main through-life cost drivers must be determined. As a basis, costs for a 16m force protection craft are used. This design is considered comparable in displacement and operational speed to the common launch and therefore provides comparable through-life costs. The main cost divers are shown to be;

- Consumables
- Operational Manpower
- Government Furnished Equipment (GFE)
- Design & Manufacture
- Upkeep planned and unplanned
- Spares

The full breakdown of these costs are shown in Figure 1

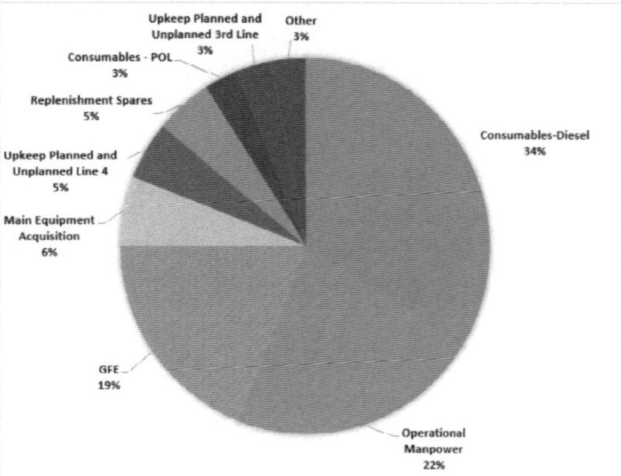

Figure 1 – Through-life cost breakdown

The top four costs make up 81% of the overall through-life cost of the vessels including; consumables, operational manpower, GFE and acquisition costs.

Herein, we therefore consider and quantify, where possible, the top four cost drivers in relation to the common launch, outlined below. The study aims to understand the overall effects on through-life costs of a common design compared to like-for-like replacement.

3.1 CONSUMABLES

The most significant through-life costs for high speed craft are consumables. Due to the common launch having to meet all requirements, a larger vessel than required is used for the majority of roles. The larger vessel has increased resistance, and therefore requires higher installed power to meet the same operational speeds. This added resistance and increased engine sizing results in increased fuel consumption. The engine and propulsion system may also be required to operate in an off design case to meet design speeds and further reducing efficiency.

Using the resistance curve from the Common Launch, installed power for each operational speed is calculated and associated fuel burn estimated. This has also been calculated for the current in-service vessels using the installed power and engine duty cycles. Total fuel burn for the current in-service vessels and the common launch to undertake a single evolution can be seen in Table 1.

Table 1- Fuel Burn Comparison

	Current Craft Fuel Burn (L)	Common Craft Fuel Burn (L)	Variation %
Weapon Training	818	1,579	93%
Dive Support	10,979	19,236	75%
Utility	1,220	2,050	68%
Crew Training	2,287	3,844	68%
Escort	1,682	2,999	78%
Police	2,099	2,887	38%
Total	**20,364**	**35,158**	**73%**

A common launch increases fuel consumption from between 38-93% from the base vessel averaging a 58% increase to undertake all operational roles.

3.2 OPERATIONAL MANPOWER

Operational manpower is the second biggest cost driver for the in service life of the vessel. Emphasis needs to be put on maintaining the minimum operator requirement when looking at combining requirements to keep operational manpower costs to a minimum.

The common launch requirement was set to allow the minimum number of operators to remain consistent with current in service vessels for all operational roles. This ensures that the cost of operational manpower would not increase.

Common designs will allow operators to be more flexible with roles allowing interoperability of crews between vessels. These savings are not guaranteed and will have to be understood on a case by case basis.

3.3 GOVERNMENT FURNISHED EQUIPMENT

GFE will be required regardless of base design. With the assumption that the same number of overall vessels is required the GFE requirement will remain constant as GFE will be required for each operational role. It's understood that there may be some common equipment that reduces the overall requirements but this is hard to quantify at pre-concept level.

3.4 MAIN EQUIPMENT ACQUISITION

Commonality of design looks to save on non-recurring costs. These costs may be slightly elevated for more complex designs but are considered to provide savings over multiple designs. Key non-recurring costs to be reduced are;

- Design Costs
- Moulding
- Survey Costs
- Class Costs
- Documentation (BR's, Safety Cases)
- Model Testing
- ITEAP (Integrated Test and Evaluation Plan)

Total costs to replace existing craft with like-for-like replacements as well as the common launch have been developed using scaling factors. Both the unit procurement costs and non-recurring costs have been included with results shown in Figure 2.

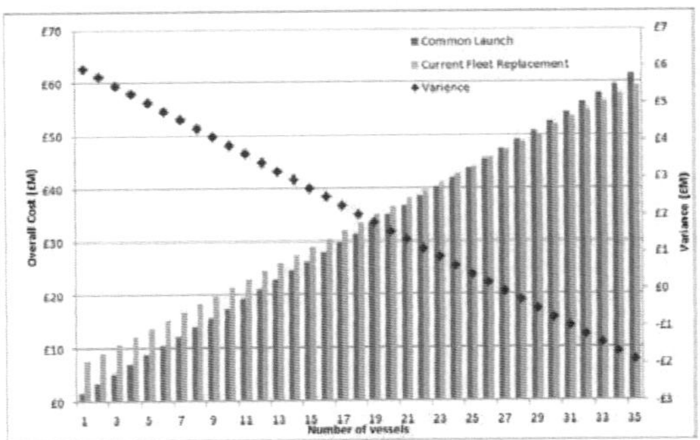

Figure 2 - Main equipment acquisition cost for a common launch and like-for –like replacement.

Due to the base vessel being larger and more complex, the non-recurring costs savings are neglected past 28 vessels. Procurement below 28 vessels will provide advantages from non-recurring costs or if fleet rationalisation allows a two vessel reduction.

3.5 COST DRIVERS CONCLUSION

There are significant cost implications to both the consumables and design and manufacture costs when looking at the common launch. Cost benefit areas are outweighed by the increase in acquisition and consumables. Results therefore show that a common design is not beneficial in terms of through-life cost and requires further breakdown of the requirement into more than one design is required.

4. RAPID DESGN TOOL - COSTING

To understand the optimum requirement combination, we use rapid design tools. Costs focus on the main cost drivers for the common launch and using the same requirements set as the common launch these are investigated to understand the most cost effective solution.

4.1 RAPID DESIGN TOOLS

The NDP have created an in-house tool HiSCAT developed using Paramarine with an embedded Microsoft Excel file.

HiSCAT was originally created to help understand the impact of varying requirements on the design of small high speed craft. It was designed specifically for the replacement of the Arctic 28 but has been adapted to fulfil new roles and requirements as they emerge. In this instance, outputs from the tool are used to quantify costs from the four main cost drivers. Firstly fuel consumption is calculated from resistance curves developed within HiSCAT to meet the operational profiles. Crewing requirements are calculated directly from the combined input matrix (See section 4.2) and rough order of magnitude costs for concept designs are calculated. Costs are developed for each design using scaling methods from known build costs including both non-recurring costs and unit procurement.

To run HiSCAT the user provides principal particulars for a vessel along with payloads, power train and basic layouts. HiSCAT then outputs an initial design spiral assessment including weight breakdown, resistance, range, stability along with a 3D image of the vessels outline design. This allows very rapid assessment of design changes and allows large ranges of requirements to be tested in short periods of time.

To be consistent all vessels have a Longitudinal Centre of Gravity at 35% of Length Water Line. All have the same propeller efficiency curves using conventional transcavitating propellers as defined by A, Molland et al, [3] and use a specific fuel consumption of 220g/Kw/h.

4.2 REQUIREMENTS COMBINATIONS

Once requirements for each operational role have been defined, requirement combinations can be established. For example, if three operational profiles are developed, these can all be combined to create one vessel, or can be separated to create three separate craft. Defining the three requirement sets as A, B and C the following permutations can be generated.

Table 2 - Requirement Combination

Vessel 1	Vessel 2	Vessel 3
A	B	C
AB	C	-
AC	B	-
BC	A	-
ABC	-	-

From three requirement sets there are five sets of vessel combinations that can be developed and seven vessels variants.

When the requirement set increases to four vessel combinations increase to 15 with 23 vessel variants. Past four variations assumptions have to be made to discount the total number of requirements as computing time exponentially increases.

For this study only three sets of requirements are analysed to prove the use of rapid design tools to assess requirement combination. The three sets of requirements from the common craft providing the largest variation are used. These include;

- A - Escort
- B - Patrol
- C - Police

To create the combined requirement set for the seven vessel variations initially all the role requirements are input into a matrix. The greatest values are taken for each requirement combination, setting the minimum requirement for each design and creating individual requirement sets for each of the seven variants. These are imported to HiSCAT where a high-level concept will be developed meeting the requirements for each variant.

Fixed requirements are defined such as length, beam, payload, pax and duty cycles in the initial model set up. Dependant variables such as range and speed, which are dependent on resistance and installed power, are automatically calculated using goal based algorithms. Once the requirements are met, vessel outputs are provided in a summary report.

5. RESULTS

Seven sets of requirements have been input to HiSCAT producing seven concept designs and associated costs. An example output can be seen in Figure 3.

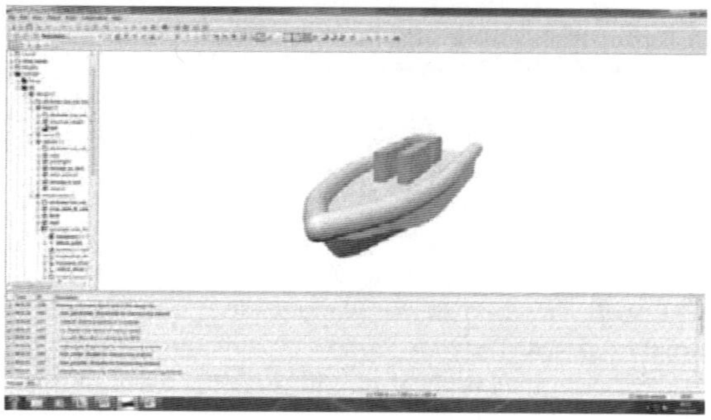

Figure 3 - HiSCAT Outputs

We have assumed that the same total numbers of vessels are required to keep the assessment consistent as it can't be guaranteed that fleet reductions can be made.

5.1 CONSUMABLE COSTS

HiSCAT outputs provide total fuel used from a single operational profile been conducted by each vessel. Estimates have been calculated from engine rating, endurance and resistance. Values can be seen in Table 3

Table 3 - Fuel burn per evolution

Combination	Vessel 1 (L)	Vessel 2 (L)	Vessel 3 (L)	Total (L)
A, B, C	6,867	22,141	3,249	32,257
AB,C	40,159	6,867		47,026
AC,B	29,522	3,249		32,770
A,BC	15,633	22,141		37,774
ABC	68,709			68,709

5.2 CREW REQUIREMENTS

Crewing position estimates have been predicted to man the full fleet assuming the maximum number of occupants required from each base vessel. Results are shown in Table 4.

Table 4 - Crew position requirements

Combination	Vessel 1 (Crew)	Vessel 2 (Crew)	Vessel 3 (Crew)	Total (Crew)
A, B, C	16	36	10	62
AB,C	70	16	-	86
AC,B	64	10	-	74
A,BC	30	36	-	66
ABC	90	-	-	90

5.3 PROCUREMENT COSTS

Costs for all vessel variants are calculated to baseline the assessment for like-for-like replacement. Results are shown in Table 5.

Table 5 - Cost per vessel

Combination	Cost Per Vessel (£M)	Number of vessels	Total
A	£2.7M	4	£10.81M
B	£1.86M	12	£22.28M
C	£2.88M	2	£5.75M
AB	£2.21M	16	£35.38M
AC	£2.49M	6	£14.92M
BC	£1.99M	14	£27.84M
ABC	£2.27M	18	£40.78M

Costs from the seven different vessel variations are used to input the vessel combinations to allow overall costs to be assessed. Combined costs can be seen in Table 6.

Table 6 - Costing for three requirement combinations

Combination	Vessel 1	Vessel 2	Vessel 3	Total
A, B, C	£10.81M	£22.28M	£5.75M	£38.84M
AB,C	£35.38M	£5.75M		£41.13M
AC,B	£14.92M	£22.28M		£37.20M
A,BC	£10.81M	£27.84M		£38.65M
ABC	£40.78M			£40.78M

6. DISCUSSION

The outputs from HiSCAT have produced designs for each combination of the requirements providing consumable, crewing and acquisition costs for a replacement fleet.

Results show that a like-for-like replacement of each vessel (A,B,C) provides the lowest consumable cost, rather than a single combined vessel. This is expected as each design meets the minimum requirement reducing excess equipment, storage and fuel. This in turn reduces the resistance, and therefore the installed power. Combination three, only combining requirements for vessels A and C produces a fuel increase of only 2% whilst combining all the requirements creates an overall consumables increase of 113% for a single evolution.

The crew output for the common craft shows an increase in the overall crew positions requirement from 62 to 90, a 31% increase. To discount this as a factor the requirement needs to be set to allow minimum crewing requirement across all vessels. This would therefore limit the crew requirement to 62 resulting in neutral cost benefits.

Procurement costs again show combining all the requirements (ABC) produce higher overall costs. The optimum output requires two vessels, with requirements A and C being combined to a single vessel. This provides cost savings of 9% compared to combining all requirements (ABC) and a 4% saving if replacing the full fleet with bespoke designs (A,B,C).

Only the top four cost drivers have been included within this analysis. Other cost saving from spares, maintenance and training will provide smaller savings when compared to the main cost drivers but still warrant consideration.

Not included with the main acquisition cost is the process of running a tender. Procuring a new vessel has significant resource and funding requirements. A standard tendering process would typically consist of;

- Writing and evaluating;
  - requests for information,
  - pre-qualification questionnaires,
  - invitations to tender,
- Contract negotiation and awards,
- Funding business cases for tenders and purchases,
- Answering clarification questions,
- Attending Industry days.

For simple designs, some steps can be removed but there are still cost implications. With an increase in the complexity of a common craft the time to write the requirement and overall assessment will increase. Resourcing to run six separate procurements may therefore not be feasible, which may benefit combining requirements.

Furthermore, with common systems, a number of cost savings can be made through the life of a vessel by combining spares, crew training and vessel maintenance. Such benefits through-life are harder to quantify, however are all factors that can contribute to savings within a common design. Further investigation is required to understand the trade-offs between cost savings of separate vessels, and the practicality and feasibility of combining requirements

7. CONCLUSIONS

This study demonstrates the use of a novel rapid design tool to understand the costs of combining vessel requirements at an early design stage. Within industry, it is often considered beneficial to combine vessel requirements into common designs. However here, we demonstrate that in the case of a common launch, combining police, escort and patrol requirements for multiple operations into a single vessel is the least viable option. Rather, outputs from HiSCAT, our rapid design tool, show that the most cost-effective solution in this case is to combine requirements for A- Escort and C-Police, whilst keeping B -Patrol as a separate design.

HiSCAT is a high-level assessment to provide a method for quantifying costs of combined requirements. In this study, we ran evaluations based on the top four cost drivers, which together form 81% of the through-life costs of a vessel. Further work will benefit understanding of the trade-offs between cost savings of separate vessels, and the practicality and feasibility of combining procurement, training and maintenance costs that can be achieved through combining requirements.

8. ACKNOWLEDGEMENTS

I would like to extend thanks to Robert Hayes from Frazer-Nash Consultancy for brain storming and pre-work conducted on cost saving areas and Alice Trevail for proof reading and support.

9. REFERENCES

1. Molland, A. Turnock, S. Hudson, D., 'Ship Resistance and Propulsion, *Cambridge University Press, 2011*

2. Atlas Electronic UK Group, SEA Class Marine Craft, *https://www.uk.atlas-elektronik com/solutions/surface-vessel-systems/sea-class-marine-craft.html, 2018*

3. S Lee, 'Common Launch Study', *Naval design Partnering, 2016*

10. AUTHORS BIOGRAPHY

Oliver Short holds the current position of Naval Architect at the Naval Design Partnering Team. He is responsible for all aspect of high speed craft from pre concept evaluation to trials and assessment.

Steven Lee holds the current position of Senior Naval Architect at the Naval Design Partnering Team. He is responsible for developing and guiding small high speed craft design and procurement ensuring requirements are met and risk areas are identified.

Tom Coe holds the current position of small boat team lead at the Naval Design Partnering Team. Tom is also a leading expert in Whole Body Vibration and conducts extensive work with the MoD ensuring all exposure to personnel is minimised.

FUTURE AIR DEFENCE: COUNTERING EMERGING THREATS

L Griffiths, BMT, Bath, UK

J Rigby, BMT, Bath, UK

SUMMARY

The relentless advance of technology forces warship designers to continually adapt and improve warship design in order to remain capable and credible against modern adversaries. The emergence of fast, accurate detection and targeting systems coupled with hypersonic and anti-ship ballistic missiles will force ships to adapt in order to survive future force level engagements. This paper analyses a mixture of capabilities and assets that can help warships to fight and survive in future battlefields. The research behind the paper investigates the effectiveness of using remote sensors deployed on UAVs to increase the detection range of the vessel, as well as the manpower and platform requirements for such a vessel. In conclusion, these remote sensors can be deployed with great effect increasing the probability of interception. However, these solutions come with reasonable manpower and platform space requirements that could be easily integrated into a new platform but would present significant challenges if integrated into the existing fleet.

GLOSSARY

AEW	Airborne Early Warning
ASM	Anti-Ship Missile
ASV	Autonomous Surface Vessel
ASW	Anti-Submarine Warfare
CEC	Common Engagement Capability
ECM	Electronic Countermeasures
EW	Electronic Warfare
MALE	Medium Altitude Long Endurance
TERN	Tactically Exploited Reconnaissance Node
UAV	Unmanned Aerial Vehicle
UxV	Unmanned vehicle (of any type)
VTOL	Vertical Take-Off and Landing

1. INTRODUCTION

Advances in semiconductor and specifically, mobile technology over the last two decades have led to a step change in the way warships communicate, coordinate and fight in the naval battlespace. However, these advances have also been exploited in the creation of ever more sophisticated anti-ship missiles (ASMs).

ASMs fielded in recent decades have been epitomised by high subsonic guided weapons such as the French Exocet missile. While supersonic ASMs such as the P700 Granit [1] were fielded, these were limited to the Soviet and Russian Navies.

This situation has changed with the introduction of supersonic ASMs such as BrahMos and the efforts of a number of nations to develop and field hypersonic ASMs in meaningful numbers. Given the increasing prevalence of these high speed weapons, future warships will have to be equipped to counter this threat.

Faster missiles reduce the reaction time available for defending warships and their increased kinetic energy means that the safe range at which an interception can take place before debris and shrapnel hit the ship also rises as shown in Table 1.

Table 1 - Time from crossing 30m mast radar horizon to impact and safe intercept assuming ballistic trajectory after intercept

ASM speed at 5m above sea level	Reaction Time (seconds)	Safe Intercept Range (km)
Mach 0.95	60.628	0.326
Mach 8	7.200	2.749
Difference	-53.428	+2.423

In order to combat these threats the design of future warships must evolve. To counter the speed of these weapons either Reaction Times or Detection Ranges or both must improve.

2. SHORTENING REACTION TIMES

Shortening reaction times is a non-trivial exercise, combat systems are optimised to detect, identify and react to incoming threats in the shortest possible time. This permeates all aspects of the system design from cabling length to operations room layout. Automatic engagement in high risk environments where the ship detects, assesses and acts without human intervention would remove the human element of the firing chain reaction time. Human-in-the-loop systems would still be possible but would need to utilise an override/abort authority in contrast to fire on command systems.

There are other methods of improving reaction times and reducing the susceptibility of a warship. High fidelity operational analysis tools can conceivably be used in near-real time to optimise a ship's air defence posture. This would draw on data from the ship combat, platform management and bridge systems held within a common computing environment.

The exploitation of quantum computers in the naval domain could allow the fielding of Quantum Radar [2] which has the potential to see through noise and clutter. However, this advance is unlikely to see implementation in the near future.

Given the difference in reaction time between sub-sonic and hypersonic weapons is so large and that only marginal gains can be obtained without quantum computing, it was concluded that shortening reaction times on its own could not create a viable solution. In order to counter the threat of hypersonic missiles the detection radius must also be increased.

3. INCREASING DETECTION RADIUS

Due to the curvature of the earth, no matter how powerful the warship radar is there will always be a set distance at which surface objects (or sea skimming missiles) can be detected. This distance or "radar horizon" can be estimated using Equation 1.

$$D_h = \sqrt{2 \times H \times \left(\frac{4R_e}{3}\right)} \qquad \text{Eqn. 1 [5]}$$

Where D_h is the radar horizon in metres, H is the height of the radar in metres and Re is the radius of the earth, normally taken to be 6,400 km.

Using a standard warship mast height of 30m the radar horizon can be estimated as 23km (not accounting for wave height oscillation). However, simply designing ships with a taller mast does not solve the problem. Doubling the height of a mast from 30m to 60m only increases the radar horizon 41%, at the cost of a significant impact on the platform design and stability. Instead, offboard sensors may prove to be a more viable solution.

In a task force, ship radar range shortcomings can be resolved through the use of airborne early warning (AEW) using either fixed or rotary wing assets. However, this is a constrained solution; it requires a number of manned aircraft to provide persistent surveillance and a platform capable of launching and maintaining those aircraft. A US carrier task group will normally embark 4-5 fixed wing AEW planes to provide persistent coverage.

If an Air Warfare ship is to provide a credible organic capability then future designs may need to be designed with space for embarked AEW. If traditional fixed or rotary wing assets were employed this would result in a much larger vessel than current destroyers and cruisers, likely reducing the number able to be procured for an equivalent budget.

Alternatives to large manned AEW aircraft do exist; an Aerostat for instance could be accommodated on a reasonably sized vessel [3] although more than one would become too onerous on the ship arrangement and performance. Such a design was suggested for the Dreadnought 2050 [4] concept, however, once maintenance, the impact on other ship operations and an inability to manoeuver are considered this option loses its attraction.

Autonomous Surface Vessels (ASVs), while promising in some roles, are ill suited in supporting anti-air warfare. To support a ship in transit they would need to be capable of keeping pace with the mothership even in heavy seas, while carrying a sensor powerful and high enough to provide the mothership with actionable data. Such a specification would drive the size of the ASV and mothership up. Semi-submersible ASVs do go some way to resolving this issue but still drive the mothership size upward.

A further possibility is the use of Unmanned Air Vehicles (UAV) which can provide a solution far closer to that of existing AEW aircraft. UAVs can achieve the necessary height and transit speed at the sacrifice of endurance.

By using a UAV positioned directly above the vessel (as presented in Figure 1), assuming a reasonable cruising altitude of 5,000m, this range can be increased to nearly 300km, providing a significantly enhanced surface radar horizon and increasing the time to react to an incoming missile threat.

This detection range can be increased further by moving the UAV to a forward detection station, increasing the radar horizon by the same distance of this forward operating position. This setup has the additional benefit of making the mother ship harder to position by the enemy (compared to it being directly below the visible UAV).

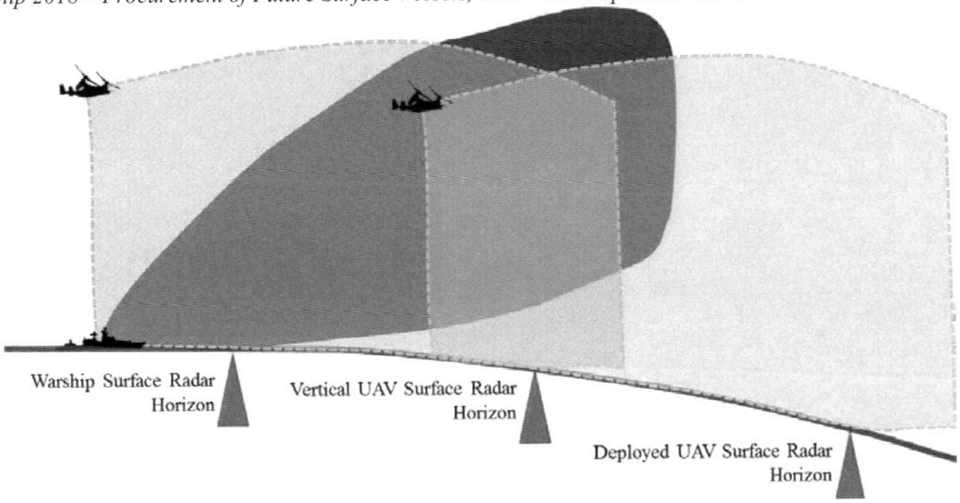

Figure 1 – Diagrammatic effect of using a UAV to increase effective radar horizon.

An important consideration to factor in when looking at radar detection ranges is the radar power and detection range. This can be determined by the radar range equation presented in Equation 2.

$$\boldsymbol{R_{max}} = \sqrt[4]{\frac{P_S\, G^2\, \lambda^2\, \sigma}{P_{Emin}\,(4\pi)^3\, L}} \qquad \text{Eqn. 2 [5]}$$

Where R_{max} is Maximum Range, P_S is Transmitted Power, P_{Emin} is Return Power, G is Antenna Gain, λ is Wavelength, σ is Radar cross section and L is Loss factor.

All of these factors play in together, along with environmental effects such as attenuation, rain, multipath effects, evaporation duct effect and surface-based duct effect to create the final effectiveness of the radar system. An example of a Radar Coverage Diagram showing percentage of detection probability is presented in Figure 2. This shows that, even if theoretically the surface radar horizon is significantly increased, the radar coverage may still limit detection range. These calculations and environmental factors were taken into consideration for our later analysis and modelling. This raises the important point that any air vehicle will need to transmit with considerable power to achieve the required range. The size and power source of the radar would need to be considered.

It is also noted that using AEW Systems is only part of the doctrine of "layered defence", it is also common to place Air Warfare ships up threat from the capital ships with further outer layers to provide early warning of launch (possibly using submarines) and also seeking to head off the threat at source (using pre-emptive attacks/special forces). This paper focuses on the up threat role of the Air Warfare ship or a lone operating ship and aims to ensure that it is survivable when faced with a hypersonic missile threat.

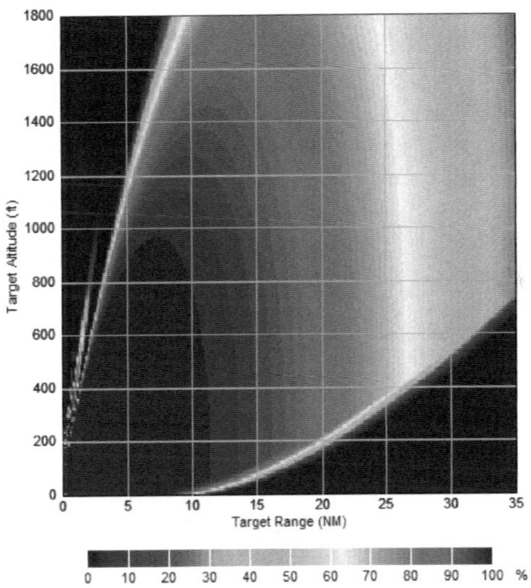

Figure 2- Example of a Radar Coverage Diagram showing percentage of detection probability

4. OFFBOARD SENSOR IMPACT

If future warships are to be designed utilising offboard UAV mounted sensor systems they must provide a tangible benefit over the existing baseline. To determine the extent of this benefit we considered a number of different configurations and their impact on defence against a hypersonic ASM. An operational analysis tool was used to compute the improvement in intercept ranges.

A scenario was devised where a baseline ship with radar mounted 32.5m above the waterline would be attacked by sea skimming weapons in an open ocean scenario i.e. no terrain obstructing radar performance.

Table 2 - Results of ASM scenario analysis showing intercept range against a range of missile threats

Scenario	Description	Number of ASMs	Detection Range (km)	Intercept Range (km)		
				Mach 0.95 (NSM)	Mach 3 (BrahMos)	Mach 8 (Zircon)
1	Baseline	1	28.3	22.6	17.0	9.4
2	Aerostat	1	168.0	134.4	100.8	56.0
3	UAV 0km	1	196.3	157.0	117.8	71.7
4	UAV 50km	1	249.9	199.9	149.9	89.6
5	UAV 100km	1	298.7	240.0	180.0	95.0
6	UAV 200km	1	398.0	318.4	238.8	133.0
7	2xUAVs 50km	8	Min: 194.0 Max: 249.9	Min: 155.0 Max: 199.9	Min: 116.3 Max: 149.9	Min: 68.0 Max: 89.6
8	2xUAVs 100km	8	Min: 174.0 Max: 300.0	Min: 139.2 Max 240.0	Min: 104.2 Max: 180.0	Min: 59.0 Max: 95.0

The three weapons chosen would attack on a direct path at sea skimming altitude (5m) with speeds of Mach 0.85, Mach 3 and Mach 8. This is roughly equivalent to a high-subsonic ASM such as Kongsberg's Naval Strike Missile [6], a supersonic ASM such as BrahMos [1] and a hypersonic ASM such as Zircon [7] respectively.

The time and distance at which the baseline ship managed to intercept the missiles was then compared against the following systems:

- Baseline + Aerostat above the ship;
- Baseline + UAV above the ship;
- Baseline + UAV 50km from the ship;
- Baseline + UAV 100km from the ship;
- Baseline + UAV 200km from the ship.

The analysis made a number of simplifications:

- The ASM was assumed to be detected upon entering radar range;
- UAVs were expected to be on the flight path of the incoming missile;
- The ASMs were expected to travel on a straight path with a 100% hit probability unless intercepted by an air defence missile;
- The interceptor was assumed to have a 1.0 hit probability against the ASM;
- All countermeasures including soft-kill, hard-kill and electronic countermeasures (ECM) were ignored;
- Reaction times on the ship were assumed to be instant;
- The launch frequency of interceptor missiles was assumed to be 1 per second.

As shown in Table 2 the baseline vessel was capable of intercepting a hypersonic missile but only if the assumptions stated above held true. Only 9.4km or 3.5 seconds separates the missile from the ship. A further analysis examined an imperfect system where a delay of 5 seconds and an imperfect intercept probability confirmed that the baseline vessel would not be able to intercept a hypersonic ASM if it relied on its own radar.

Against a Mach 8 target, an aerostat or UAV directly above the platform would be sufficient to extend the intercept range beyond the ship's radar horizon. Placing the UAV up the threat axis again pushes out the engagement envelope even further.

A further 2 scenarios were considered with multiple UAVs and missiles in order to stretch the limits of the proposed solutions and assess whether the benefits were still worthwhile in scenarios where the missile approached from a non-optimal direction:

- Baseline + 2 UAVs 50km from the ship;
- Baseline + 2 UAVs 100km from the ship.

In this scenario the UAVs were placed on opposite sides of the ship as illustrated in Figure 3. This scenario showed that even when the UAV is on the wrong side of the ship its height still allows for the detection and interception of a missile before entering the ship's radar horizon.

Against slower moving targets and those relying on low observable features, this additional range allows for more thorough interrogation of false tracks and a larger window in which to detect missed tracks.

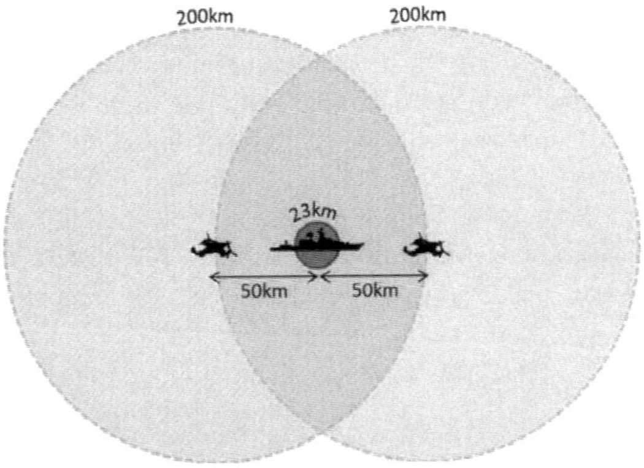

Figure 3 - Representation of the increased surface radar range with the addition of two UAVs at a distance of 50km from the ship.

Increasing the detection radius also gives the commander enough time to have the option of firing a single interceptor knowing that if the intercept fails there is still time to fire another.

5. FEASIBILITY OF AIRCRAFT OPERATIONS

It is important to consider the feasibility of such an arrangement with overboard UAV sensors. The first step was to look at the available types of aircraft.

5.1 TYPES OF AIRCRAFT

By limiting the aircraft considered to those able to launch from a typical helipad or deck space on a non-carrier type warship, three main types of aircraft were considered:

- **Small Fixed Wing Aircraft**
  Launched by catapult or similar located on the flight deck. Typically has longer endurance, but a smaller payload than a rotary winged UAV.
- **Rotary Wing Aircraft**
  These Vertical Take-off and Landing (VTOL) Aircraft are well suited to the limited space available on a ship's deck, but often as a result have limited endurance and range.
- **Tiltrotor – VTOL**
  These aircraft combine the endurance and speed of fixed wings and the VTOL capability of rotary wings. Performance in flight is normally equivalent to fixed wing aircraft.

For the purposes of the later analysis we have selected example aircraft of the three different types in order to determine the key characteristics as presented in Table 3.

These examples show what UAVs are available now and therefore do not necessary reflect the developed capabilities that will be deployable from a future air warfare destroyer. For example, it may well be that in the future a more capable quadcopter option may be available to provide a stable unmanned VTOL platform with increased endurance.

Additionally if operations were not limited to a small flight deck, alternative systems like Boeing's Hale or the RQ-4 Global Hawk could be applied to vastly increase payload and endurance [13]. This analysis instead utilised the worst case assumption that land based or carrier based aircraft support is unavailable and the air defence ship needs to rely on its own organic capability.

When studying the comparison data in Table 3, it should also be noted that the Bell Eagle Eye was first developed in 1998, and even then managed to achieve an endurance of over 8 hours [11]. It is expected that with recent technological advances in efficiency this range could be further improved if necessary. Projects like the Bell Eagle Eye have recently received increased attention due to the DARPA Tactically Exploited Reconnaissance

Table 3 - Comparison of available aircraft types for AEW tasking

Aircraft Type	1. Small Fixed Wing	2. Rotary Wing	3. Tiltrotor
Image		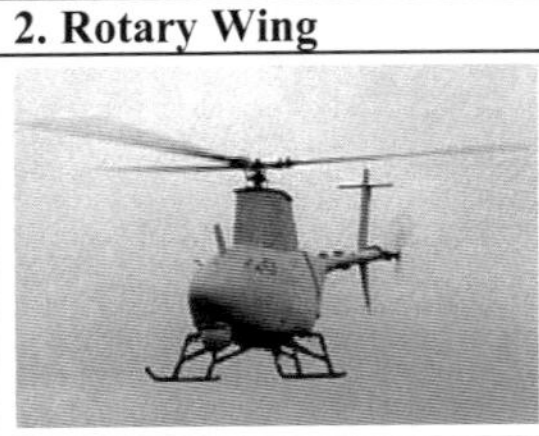	
Example	Boeing RQ-21 Blackjack	NG MQ-8B Fire Scout	Bell Eagle Eye Tiltrotor
Size	L= 2.5m Wingspan= 5m	L=7.3m Rotor Dia=8.4m	L= 5.46m Wingspan= 7.37m
Maximum Take Off Weight	61kg	1430kg	1020kg
Max Payload	20kg	272kg	90kg
Altitude	5,900m	6,100m	6,100m
Endurance	16 hours	5 Hours	8 Hours, New Variant expected to increase [10]
Reference	[8]	[9]	[11,12]

Node or "TERN" programme (Figure 4) [14]. This programme seeks to design, develop, and demonstrate a medium-altitude long-endurance (MALE) unmanned aircraft system and related technologies that enable future launch, recovery, and operations from small ships.

Figure 4 – DARPA TERN Concept UAV could provide VTOL long endurance surveillance, image reproduced from DARPA [14]

5.2 PAYLOAD

The two driving characteristics behind the vehicle selection are payload and endurance. Endurance can be supplemented by operating multiple vehicles in shifts; however payload is an unavoidable requirement. The important question to ask is how big does the radar need to be?

As a baseline, the SEARCHWATER® 2000 Radar, used for the Sea King mounted early warning system has a weight of around 900kg [15] along with the supporting consoles. By contrast the new SEARCHMASTER® radar also produced by Thales weighs only 78kg [16] enabling the Tiltrotor or the rotary wing aircraft to carry the payload. Without significant advances in fixed wing UAV launch and recovery it would still not be possible to utilise a Fixed Wing aircraft with this payload from a small flight deck. The small fixed wing UAV was therefore eliminated from the study.

Another important feature yet to be determined is if a reasonably sized UAV would be able to provide the power required to give the radar sufficient range. Due to limited information available this question was not sufficiently answered and would be a primary consideration for future work. The study was limited to considering sensors placed onboard the UAV platforms. An extension of this work would consider the feasibility and implications of also mounting effectors such as anti-air missiles on the UAVs.

5.3 MANNING REQUIREMENTS

The next aspect to consider is the human crewing requirement to maintain and operate the system. Estimates for aero engine maintenance hours per flight hour can vary from 2 to 0.4 hours [17]. The Bell Eagle Eye for example required less than 1 hour of maintenance time per flight hour [11]. The lower value of 0.4 hours maintenance per flight hour has been assumed, although it is understood that this figure will only be achieved with significant design implications.

It is assumed that the UAV would be operated autonomously requiring no input for take-off, transit, station keeping or landing. Only minimal intervention would be necessary meaning a single operator would be necessary. If full autonomy is realised the crew requirement for a detachment of UAVs providing persistent coverage could be reduced further. This does not consider management of the air picture which is already a task covered by an air defence ship's crew.

Assuming a 30 minute flight time to reach its station we have allowed for a 1 hour overlap between aircraft to ensure the full picture is permanently maintained. Figure 5 presents the analysis to determine the equipment and crew requirements in order to support 24 hour continuous surveillance.

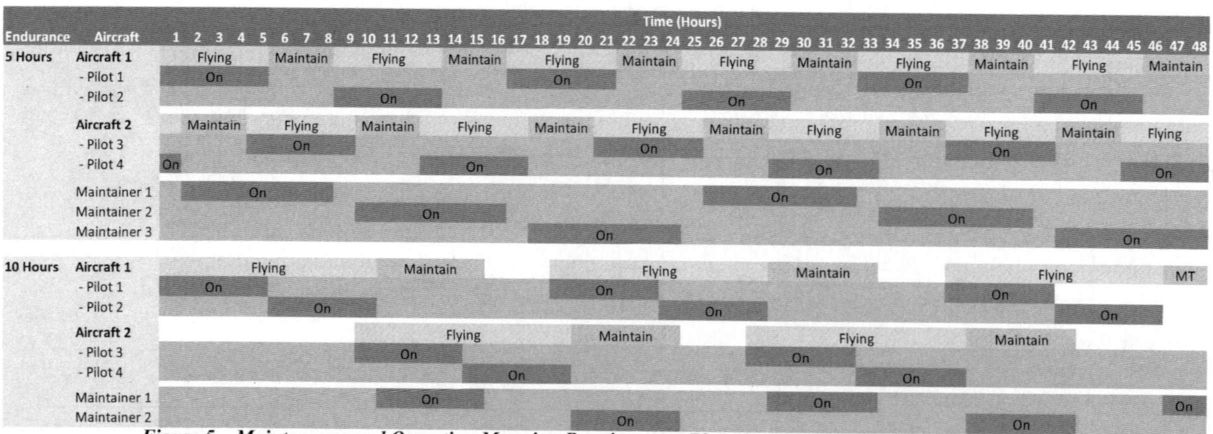

Figure 5 – Maintenance and Operation Manning Requirements Plotted over a 48hour constant watch period

5.4 AIR WING REQUIREMENT

In conclusion, depending on the aircraft endurance we would expect that for every aircraft airborne, 3 aircraft in total will be required to be embarked. Although it is possible to operate a 24 hour watch with just two aircraft, this assumes that only routine preventative maintenance is required. If this tempo is to be continued for a longer period of time the number of airframes would need to number at least 3 to factor in long term maintenance.

Assuming a reasonable level of automation and reduced maintenance to current arrangements, from the calculations visualised in Figure 5, we estimate that 3 aircraft and 7-8 crew would be required to maintain a persistent surveillance capability.

5.5 LAUNCH AND RECOVERY

A tiltrotor UAV the size of Eagle Eye would have no requirement for additional launch and recovery apparatus beyond deck handling equipment. The inclusion of a deck arrest system on Eagle Eye further enhances its ability to operate in heavy sea states where fixed wing UAV launch and recovery systems would be dangerous to operate.

5.6 STOWAGE AND MAINTENANCE ENVELOPES

Assuming 3 UAVs are required for persistent AEW a hangar of sufficient size for their stowage and maintenance will be necessary.

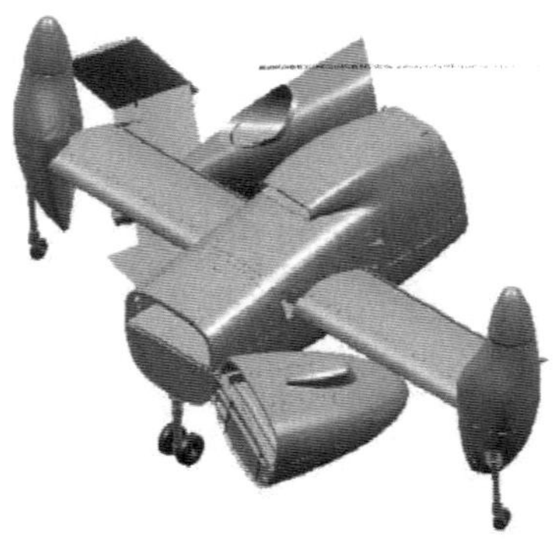

Figure 6 - Eagle Eye folded gives significant access to avionics/payload [11]

With a footprint of 7.4x5.6m when fully deployed, 3 Eagle Eye UAVs can fit into the space required to maintain one large helicopter. Folded it is possible to fit 2 Eagle Eye's in a 7.62x3x1.4m box. At 1.4m high it is also short enough that it could theoretically be stacked giving users the option to store and maintain 3 UAVs in the footprint of a light/medium weight shipborne helicopter.

5.7 COMMAND AND CONTROL

Cooperative Engagement Capability (CEC) is a US Navy system designed to allow battlegroups to share sensor data in near real time to create an integrated air picture. It enables the separation of detection, tracking, shooter and guidance air defence functions. Multiple radar pictures can also be overlaid creating a more accurate picture. AEW UAVs are ideal assets to enhance this capability with the ability to seamlessly integrate their sensor data and tracks into the battlegroup airspace picture. It should be noted that creating the aforementioned CEC system presents some significant design challenges, such as achieving the geospatial and temporal referencing accuracies, low latency (needed for supersonic threats), and the functional/dynamic safety integrity needed.

The capacity to control unmanned air and surface vessels is expected to be a common capability on future warships. This enables them to realise the operational flexibility granted by embarking unmanned assets and controlling friendly assets within a task group. The development required to achieve this capability should not be underplayed, with significant challenges to overcome in creating a comms systems able to manage multiple UAVs and integrate with the Command and Control systems. Assuming that with sufficient funding these developments can be achieved in a reasonable timeframe; they should be available in time for future Air Warfare vessels. However, significant work would need to be completed to integrate this onto existing platforms.

6. ADDITIONAL APPLICATIONS

While this paper has focused on sea-skimming hypersonic threats this does not mean that using UAVs to provide AEW is not useful against other threats. Increased sensor coverage allows more time to detect and counter subsonic ASMs exploiting stealth and super-low sea skimming approaches or swarms that would otherwise saturate defences if detected at closer range.

Equally such versatile UAVs need not be limited to the AEW role. Whether in a task group or as a singleton multiple UAVs could be used for the following roles:

- Surveillance/Reconnaissance – Using a EO sensor/radar package;
- ASW – dropping of sonobouys;

- Littoral sensor coverage – In environments where terrain limits the host platform's sensor coverage;
- Active Radar – allowing the host ship to use its sensors passively;
- EW/ECM – As an offboard effector;
- Communications node (or relay station) - particularly relevant when seeking to create mesh communication networks for ASVs;
- Target designation.

7. CONCLUSIONS

Future warships will have to defend against supersonic and hypersonic ASMs. This threat will only grow and those responsible for the maintenance of existing ships and procurement of their replacements will have to consider how to defeat these threats. While existing radars can defend against these threats the margins for error are very small. Failure to detect on the radar horizon, a slow reaction, failed launch or failure to intercept would be deadly. The use of offboard sensors gives the warship the time necessary to observe, orient, decide and act in the face of an incoming threat.

The approach to air defence discussed in this paper does not have a significant impact on a newly built future warship design. The future warship would need to have, space for tactical data links, a small number of additional crew, expanded aircraft maintenance facilities as well as additional spares and fuel. The additional costs of these areas are potentially offset by a significant increase in the effectiveness of a warship in controlling and defending its airspace.

Although these additional platform requirements can be relatively easily integrated into a future warship design, they could cause significant disruption if implemented on board an existing vessel. This additional platform requirement highlights the need for navies across the world to start experimenting with Unmanned and Offboard Vehicles today, so that there are ready to adopt these systems when they emerge.

8. ACKNOWLEDGEMENTS

The authors would like to thank all those who supported the research work upon which this Paper is based and Andrew Harris, Tim Lear, Jay Hart and Stuart Eyles for their invaluable technical guidance. In particular special thanks go to Ben Lindley-Start for his operational analysis modelling which contributed to this paper.

9. DISCLAIMER

The opinions expressed in this paper are those of the authors and do not necessarily represent BMT.

10. REFERENCES

1. C. Kopp, 'Soviet/Russia Cruise Missiles Technical Report APA-TR-2009-0805', [online] Available at: http://www.ausairpower.net/APA-Rus-Cruise-Missiles.html [Accessed 6 Aug. 2018], 2012
2. ASPI, 'From Little Things: Quantum Technologies and their application to defence.', [online] Available at: https://www.aspi.org.au/report/little-things-quantum-technologies-and-their-application-defence [Accessed 6 Aug. 2018], 2017
3. CIMSEC, 'The Strategic Role of Tactical Maritime Aerostats in Ensuring Persistent Surveillance.', [online] Available at: http://cimsec.org/strategic-role-tactical-maritime-aerostats-ensuring-persistent-surveillance/26214 [Accessed 5 Aug. 2018], 2018
4. Alan Tovey, 'Dreadnought 2050: Here's what the Navy of the future could be sailing', [online] Available at: https://www.telegraph.co.uk/finance/newsbysector/industry/defence/11833043/Dreadnought-2050-Heres-what-the-Navy-of-the-future-could-be-sailing.html [Accessed 5 Aug. 2018], 2015
5. RF Café, 'Electronic Warfare and Radar Systems Engineering Handbook - Radar Horizon / Line of Sight - RF Cafe.', [online] Rfcafe.com. Available at: http://www.rfcafe.com/references/electrical/ew-radar-handbook/radar-horizon-line-of-sight.htm [Accessed 5 Aug. 2018], 2018
6. Kongsberg, 'Naval Strike Missile', Kongsberg, 2014
7. TASS, 'Russia's hypersonic Zircon anti-ship missile reaches eight times the speed of sound', [online] Available at: http://tass.com/defense/941559 [Accessed 5 Aug. 2018], 2017
8. Boeing, 'Boeing: Autonomous Systems - Integrator.', [online] Available at: https://www.boeing.com/defense/autonomous-systems/integrator/index.page [Accessed 5 Aug. 2018], 2018
9. Northrop Grumman, 'Fire Scout', [online] Available at: http://www.northropgrumman.com/Capabilities/FireScout/Pages/default.aspx?utm_source=PrintAd&utm_medium=Redirect&utm_campaign=FireScout+Redirect [Accessed 5 Aug. 2018], 2018
10. Flightglobal, 'New search for VTOL UAVs may resurrect Bell tiltrotor', [online] Available at: https://www.flightglobal.com/news/articles/new-

search-for-vtol-uavs-may-resurrect-bell-tiltroto-421075/ [Accessed 5 Aug. 2018], 2018
11. Bell, 'Eagle Eye Pocket Guide', Bell Helicopter, 2005
12. FAS, 'Eagle Eye UAV', [online] Available at: https://fas.org/irp/program/collect/eagle-eye.htm [Accessed 5 Aug. 2018], 2018
13. Boeing, 'Boeing: Autonomous Systems - High Altitude Long Endurance (HALE)', [online] Available at: https://www.boeing.com/defense/autonomous-systems/hale/index.page [Accessed 5 Aug. 2018], 2018
14. DARPA, 'TERN', [online] Available at: https://www.darpa.mil/program/tern [Accessed 5 Aug. 2018], 2018
15. Thales, 'Searchwater ASaC | Thales Group', Thales, 2018
16. Thales, 'SEARCHMASTER | Thales Group', Thales, 2018
17. Rotor.org, 'Helicopter Operating Cost Estimates', [online] Available at: https://www.rotor.org/portals/1/membership/Guide%20for%20the%20Presentation%20of%20Helicopter%20Operating%20Cost%20Estimates.pdf [Accessed 5 Aug. 2018], 2018

11. AUTHOR'S BIOGRAPHY

L Griffiths – Is a Chartered Naval Architect working for BMT having previously studied at Strathclyde University, He is experienced in survivability and concept design studies and has worked both internally and externally on ship design innovation.

J C Rigby – A Naval Architect at BMT in Bath for over 4 years having previously trained at UCL. Specialising in the areas of Ship Signatures and Survivability, he also has experience of autonomous systems and their operability with ship designs. He is currently in the role of Research and Development lead.

THE ADVANCED TECHNOLOGY CORVETTE (RAILGUN): A CONCEPT EXPLORATION OF FUTURE WARSHIPS

R J Pawling, L Farrier and **N Bradbeer,** Department of Mechanical Engineering, UCL, UK

SUMMARY

High power electric weapons, such as high-energy lasers and the Electro-Magnetic Rail Gun (EMRG) are the focus of research for naval application. Previous work at UCL has considered the integration of these systems on large surface warships, and the Advanced Technology Korvette – Railgun (ATK-R) project is an ongoing UCL thought experiment concerning future small warships armed with electric weapons. It brings together three main areas of research; design methodology (ZEOLIT); ballistics modelling (HYPNOS); and PhD level research into Energy Storage Systems (ESS). This paper provides some background on aspects of EMRG weapons and ESS options, before describing the ATK-R design studies carried out so far. The paper concludes with some design trends identified and future areas for development, both in terms of EMRG integration with warships and UCL research.

1. INTRODUCTION

Warship propulsion is becoming increasingly electrified, with Integrated Full Electric Propulsion (IFEP) and hybrid electric propulsion being utilised on naval vessels from light frigates to aircraft carriers [1]. Previous work at UCL, presented to INEC, investigated the implications on ship configuration of various types of electric propulsion, including novel concepts such as vertically-mounted gas turbine alternators (GTA), applied to a large surface combatant.

Of specific interest in this paper is the application of the Electro Magnetic Rail Gun (EMRG), a device which utilises Lorentz forces generated by high power pulses to accelerate a projectile along a conducting barrel. Figure 1 illustrates a generic EMRG system, with the four main electrical components; the Energy Storage System (ESS), dis/charging over timescales of seconds; the Pulse Forming Network (PFN), which discharges into the conducting rails in times measured in nanoseconds; and the conducting armature that propels the projectile. A fifth, optional component is a muzzle shunt, which recovers some of the c.50% of input energy which remains in the rails after each shot.

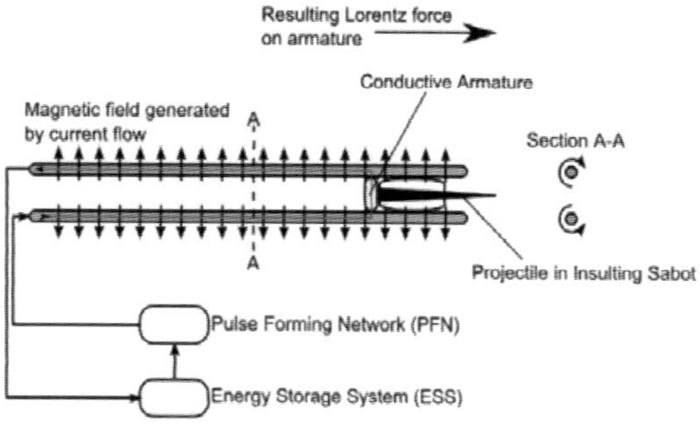

Figure 1: Block diagram of a generic EMRG [2]

The Marine Research Group (MRG) at UCL has undertaken several studies into various aspects of the integration of these future electric weapons. These have included MSc ship designs and individual projects, marine engineering research to PhD level and design studies. In his recent PhD research, Whitelegg [3] [4] investigated the transient and harmonic power system impacts of an EMRG providing 64 MJ of muzzle energy per shot, using time-domain simulation methods. This was aimed at addressing the particular problem of EMRG; that the pulsed power inputs required could adversely impact the stability of the ship power and propulsion system. Whitlegg determined that, to maintain Quality of Power Supply (QPS) for an AC distribution system, within NATO STANAG 1008 limits at a firing rate of 12 shots per minute, the ESS needed to store additional energy, a total of 305 MJ, to provide residual charge sufficient to reduce the inrush current during recharging.

Concept design studies for vessels with an all-electrical armament were presented by Andrews et al [2] and [5], illustrated in Figure 2. Armed with EMRG and Free-Electron Lasers, these vessels represented hypothetical replacements for the RN Type 45 destroyer, and were subsequently large enough to easily accommodate high electrical power generation and large ESS, as indicated by the summary of principal particulars for the original monohull design in Table 1.

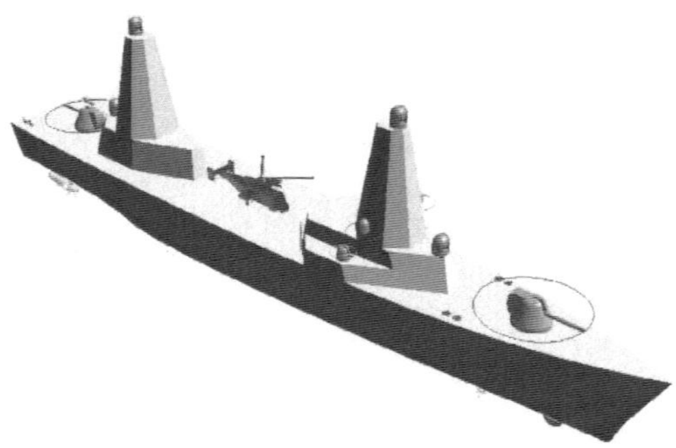

Figure 2: UCL electrically armed destroyer of 2010 [2]

Table 1: Principal particulars of the 2010 UCL all-electric destroyer monohull design

Displacement	8740 tonnes (deep)	*Installed Power*	63.2MW
Internal Volume	36500m^3	*Propulsion Type*	IFEP + pods
Length (oa)	151m	*Electric Weapons*	2 x 62.5MJ railgun
Beam (oa)	21.7m		2 x 1.5MW IR laser
Draught	5.6m		2 x 0.6MJ railgun
Max. Speed	29 knots		

2. THE ATK-R STUDY

The Advanced Technology Korvette – Railgun (ATK-R) is an ongoing UCL study being used as a focus to interface several different areas of research at UCL; Naval Architecture; Combat Systems Engineering; and Marine Engineering. ATK-R is a thought experiment around the smallest ships that can carry electric weapons, with "small ship" being defined as 500-4000 tonnes standard displacement. This paper focusses on the ship design and combat systems sides of the project, with more information on the marine engineering aspects being included in a sister paper to INEC [6]

The Naval Architectural aspects of ATK-R are focussed on methods of concept exploration, an established area of research and development at UCL, and in particular a prototype tool "ZEOLIT", described in more detail in Section 5. Combat Systems Engineering aspects are primarily in the area of EMRG sizing, ballistics and effects, with the consequences for future Concepts of Operation (CONOPS) of these weapons and their interaction with power and propulsion systems. ATK-R is making use of ongoing research work at UCL in the area of Energy Storage Systems (ESS), which is described in more detail in [7].

3. ADVANCED TECHNOLOGIES

3.1 COMBAT SYSTEMS

ATK-R contains certain implicit assumptions regarding future combat systems, such as reduced crewing requirements and lightweight multifunction radars. The two technologies explicitly examined are railguns and Uncrewed Vehicles (UxVs).

3.1 (a) EMRG

Developmental EMRG range from large calibre weapons with muzzle (projectile) energies of around 64MJ, [8] to smaller calibre anti-air weapons, such as the Blitzer prototype [9]. The potential for control over the acceleration imparted to the projectile raises the possibility of the larger-diameter EMRG being used as launchers for items such as expendable UAVs and loitering munitions, which would deploy wings to glide for some time in the target area, providing reconnaissance, battle damage assessment (BDA) etc.

Table 2 summarises the range of EMRG options of interest in ATK-R. The physical size of the weapons has been estimated using the models previously used in UCL studies. Maximum in-bore acceleration was limited to approximately 25,000G based on claimed capabilities for the Oto-Melara Vulcano 127mm projectile [10]. The columns to the right of Table 2 illustrate some of the contrasts of the EMRG; the instantaneous power in the breech is measured in GW, but the total energy used for each shot is tiny compared to that routinely consumed for propulsion and hotel services.

Table 2: The range of EMRG sizes under investigation

Type	Projectile Diameter	MV	Projectile Mass	Bore Diameter	Muzzle E	Barrel Length	Launch Mass	Mount Mass	Aux Pwr Excl Cooling	Average Accel	Breech NRG	Breech Pwr	Breech Energy
	mm	ms-1	kg	mm	MJ	m	kg	te	kW	G	MJ	GW	MWhrs
Heavy 1	127	2000	18.18	157	55.72	10	27.86	30.00	180.00	20387	142.88	14.29	0.0397
Heavy 2	97	2000	12.73	127	37.10	9	18.55	17.66	105.97	22653	95.14	10.57	0.0264
Medium HV	70	2000	8.00	100	22.37	8	11.18	9.71	58.28	25484	57.35	7.17	0.0159
Medium LV	70	1500	8.00	100	12.58	8	11.18	9.71	58.28	14335	32.26	3.02	0.0090
Light	57	1500	3.90	87	6.85	7	6.09	6.49	38.95	16383	17.58	1.88	0.0049
UAV	127	1000	10	157	9.84	10	19.68	30.00	180.00	5097	25.23	1.26	0.0070

For the larger EMRG dedicated to proving long-range fires, a semi-submerged configuration was developed with the barrel limited to >20 degrees elevation, as illustrated in Figure 3. This trades off increased internal volume demand and minimum range for a lower centre of gravity and less obstruction of the bridge sightline. Integrating external barrel supports may also be necessary if very long barrels are required for maximum muzzle velocity.

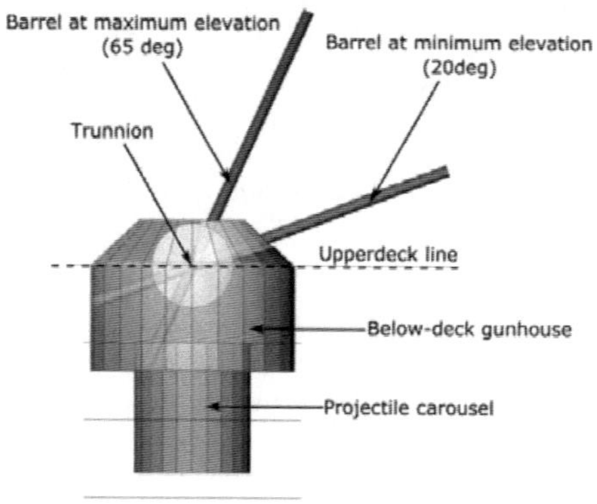

Figure 3: Concept for a semi-submerged EMRG configuration for larger calibre weapon

3.1 (b) UxVs

The larger ATK-R variants are intended to make use of a number of Uncrewed Vehicles, specifically Aerial Vehicles (UAVs) to provide over-the horizon detection and targeting for the EMRG. Two main types were considered to be of interest; a gun-launched expendable UAV, deploying wings for a prolonged glide; and a small rotary-wing UAV, capable of carrying a larger sensor system.

The rotary-wing UAV is a notional design based on the Bombardier CL-327 / CL-427 designs, which entered the prototype stage in the late 1990's. This design was chosen as it utilises a co-axial contra-rotating rotor arrangement, permitting a very compact vehicle. The principal particulars of this UAV are shown in Table 3.

Rotor Diameter	4m
Height	1.85m
Footprint	0.6m x 0.6m
MTOW	234kg
Payload	105kg
Maximum speed	85knots
Endurance	5hours
Hours at 100km radius	3
Ceiling	5500m

Table 3: Principal particulars of the CL-327 UAV [11]

Although most modern rotary-wing UAVs have a conventional tail-rotor arrangement, coaxial options do exist, and the CL-327 was developed into the German SEAMOS drone. This configuration has advantages for use on smaller vessels, in that, if the rotor blades can be removed or folded down along the body, rather than to the rear as in tail rotor arrangements, the stowage footprint is very small.

A key issue that arises when considering extensive use of UAVs on ships is the crewing requirements for handling and maintenance. Given the growth in civilian and government use of smaller UAVs, those vehicles with strictly limited capabilities – such as those used in ATK-R – would be expected to reduce in unit cost as the technology develops. The baseline option adopted for ATK-R was a wing of 10 UAVs with an automated handling system scaled from recent automated magazine systems, shown schematically in Figure 4.

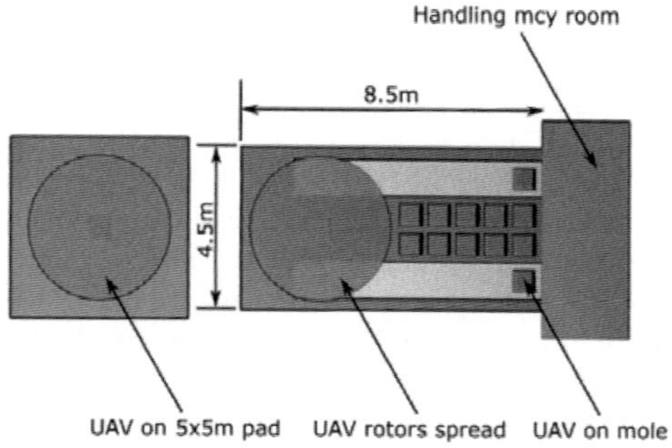

Figure 4: Indicative schematic of automated UAV handling and stowage system

The use of a highly automated handling system introduces a problem space where the requirements of light weight, low cost and high reliability may be in conflict with each other. For the initial ATK-R study, the handling system had approximately twice the mass of the UAVs it carried, based on estimates of mass for automated ammunition handling systems.

4. COMBAT SYSTEMS CONCEPTS

4.1 EMRG CAPABILITIES

The different calibre EMRG were sized for different roles; land attack, Anti Surface Vessel Warfare (ASVW) and Anti-Air Warfare (AAW) against Anti-Shipping Missiles (AShM) or Counter Rockets, Artillery and Mortar (CRAM). These were primarily based on those roles associated with recent developments in guided and course-corrected projectiles, which offer the greatest potential to utilise the EMRG capabilities.

Table 4 provides an overview of the indicative scenarios used to size ESS for the EMRG options. Smaller calibre weapons were sized for scenarios with multiple engagements each involving firing a burst of rounds. The ATK-R project has also involved the development of a hypersonic external ballistics model, HYPNOS, to evaluate the performance of the various EMRG options.

Name	Type	Calibre	Rounds per Engagement	RPM	Engagements per Scenario	Scenario Duration	Interval Between Engagements
		mm				s	s
ASVW 1	Continuous	127	20	16	1	150	75
ASVW 1	Continuous	97	20	16	1	150	75
AAW-AShM	Burst	70	3	120	4	60	13.5
ASVW 2	Burst	70	10	80	4	120	22.5
AAW-CRAM	Burst	57	5	80	4	60	11.25

Table 4: Engagement scenarios initially under investigation

As guided sub-calibre projectiles are being used, the concept of "performance" is complicated beyond the conventional aspect of minimum and maximum range, for example:

- Time to range: This is significant when dealing with moving targets. Although the EMRG projectiles are hypersonic, their arcing trajectories will increase time of flight. Unplanned mid-course corrections would require a command uplink (in a crowded projectile body) and expend the finite kinetic energy of the round. The desire to achieve simultaneous time-on-target would reduce effective range below the ballistic maximum.
- Descent velocity and angle; projectiles with a seeker head must acquire the target as they descend, and manoeuvre to strike it. Seekers have a finite size and thus range, limited by the projectile diameter, and the projectile has limited manoeuvre capability. Thus there will be some trade-off between high descent velocities that both complicate interception and increase impact energy; and having sufficient time for the projectile to acquire and guide to target. Table 5 summarises the guidance methods identified as possible for the different calibres, based on a literature survey.

Table 5: Possible guidance system methods for various projectile diameters

Guidance	Minimum Projectile Diameter	Example Current Projects / Applications
Inertial + GPS	70mm	Crit
Millimetric Radar[2]	127mm (<110mm)	DASH (SSW)
Imaging Infra-Red[1,2]	97-127mm	RAM
Infra-Red[1]	57mm	ORKA
Semi-Active Laser[3]	70mm	APKWS
Command Guidance[4]	<57mm	EAPS
Beam-Riding[4]	57mm	DART

Notes:
1: Requires thin-film cooling from nose cap
2: Can identify target type, SSW may be target type-limited
3: Difficult to integrate with hypersonic nose shapes
4: Requires transmitters on launching ship

- Terminal velocity and angle; the lethality of EMRG are being investigated in a separate UCL MSc project, but one advantage associated with EMRG is the use of kinetic energy as a destructive mechanism, avoiding the carriage of explosives on the launching ship. However, manoeuvring to correct for errors and a moving target will consume this kinetic energy, and the general assumption of a steeply-diving terminal trajectory may not hold against fast-moving targets.

4.2 HYPNOS TRAJECTORY MODELLING

4.2(a) The HYPNOS Tool

A code was developed in Python to model the trajectory of EMRG projectiles, called HYPNOS (HYPersonic Naval Ordnance Simulator). This was a two-dimensional second-order time-domain numerical simulation resolving the forces on the projectile over the time of flight.

For an unguided ballistic projectile, small yaw angles were assumed, in line with approximations suggested by McCoy [13]. For simplicity, the code neglected the curvature of the earth and Coriolis force, and did not model heating of the projectile. The projectile flew a ballistic path influenced only by gravity and air drag. The Mach number and atmospheric pressure were derived at each timestep, using curve fits to the 1962 US Atmosphere Model [12]

Two projectile drag profiles were used. The first (standard model) was based on curve fitting to drag data for the Ballistic Research Laboratory's No.8 Projectile [13], an ogive/cylinder geometry. The drag coefficient CD_0 was modelled against Mach number as a constant when subsonic, a sinusoidal curve with a short constant section when transonic, and a cubic curve asymptotic to 0.175 over the supersonic and hypersonic regions. Since slender cone hypersonic geometries can achieve lower CD_0 values of 0.08 – 0.15 over the critical speed range between Mach 4 and Mach 7 [14] a second (reduced) drag profile was produced to better match those CD_0 values. The reduced drag profile was produced by uniform reduction from the first and is not conservative in the subsonic regime, considered acceptable since the projectiles did not spend any appreciable time subsonic. These two profiles are shown in Figure 5.

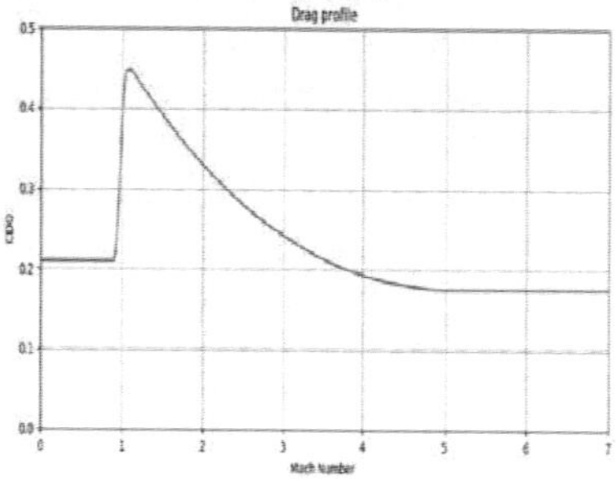

Drag profile for ogive – cylinder geometry projectile

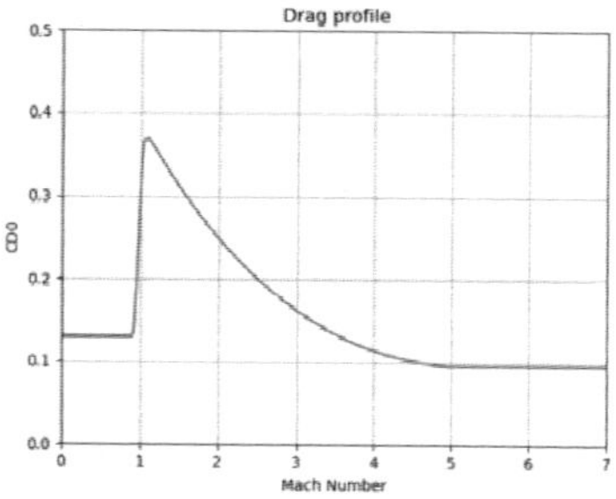

Drag profile for cone geometry projectile

Figure 5: Hypersonic drag profiles for projectiles

Since a number of assumptions (particularly relating to projectile shape and the resulting drag profile) had been made, the model was validated against data published by McNab et al [15] predicting railgun projectile trajectories. For a range of projectiles from 5kg to 60kg using the standard drag model HYPNOS predicted the flight distance with errors between -2.0% to +4.6%, which was considered to give sufficient confidence for an exploratory model.

4.2(b) Results

HYPNOS was then used to generate plots of trajectory, velocity and drag force for the range of projectiles, muzzle velocities and launch geometries under consideration. Table 6 compares the range an impact velocity for two projectile types, over a range of launch elevations, for the EMRG options in Table 2, with higher values shaded darker red. The comparison clearly shows the significance of CD_0 and thus projectile shape. This will have implications for the type of guidance that can be selected from Table 5, as some sensors will not be compatible with low-drag hypersonic geometries.

Table 6: Comparison of range and impact velocity

		STANDARD DRAG MODEL						REDUCED DRAG MODEL					
		Heavy1	Heavy2	Med HV	Med LV	Light	UAV	Heavy1	Heavy2	Med HV	Med LV	Light	UAV
Dia	mm	127	97	70	70	57	127	127	97	70	70	57	127
MV	m/s	2000	2000	2000	1500	1500	1000	2000	2000	2000	1500	1500	1000
Pr Mass	kg	18.18	12.73	8	8	3.5	10	18.18	12.73	8	8	3.9	10
	Elevation												
Range	10	27252	31016	35296	27312	22664	10709	41366	46405	51935	37779	32349	15032
(m)	20	37774	44373	52264	39472	31369	13995	64121	74497	86364	59538	48837	21045
	30	49859	61434	76161	51768	39355	15912	99188	119328	131570	84764	66964	25175
	40	67650	86879	110336	66628	48119	16877	143700	169868	196348	111795	88413	28038
	50	84939	108467	134633	78042	55816	16783	169092	194840	220155	125174	101855	29164
	60	88352	110294	133842	77202	56297	15236	163974	186007	207328	117847	98123	27543
	70	72703	89368	106945	61588	45844	11869	129082	145046	160355	91093	76907	21847
	80	40906	49861	59219	34066	25667	6608	70899	79264	87239	49537	42135	12152
V impact	10	357	398	445	442	372	261	512	569	634	612	522	362
(m/s)	20	347	387	434	406	342	229	506	569	643	570	481	311
	30	390	442	508	448	373	241	613	708	817	652	540	325
	40	449	523	617	515	418	258	758	874	995	767	633	352
	50	514	610	722	584	466	274	877	996	1115	857	718	379
	60	575	682	802	640	509	287	961	1078	1194	916	779	404
	70	620	734	857	679	542	296	1015	1132	1243	952	819	422
	80	648	765	889	702	562	302	1047	1161	1271	973	842	434

Considering only the reduced drag model, Table 6 shows that the 70mm medium high velocity EMRG offers the greatest range and impact velocity. However, to consider a unitary kinetic impactor, with no explosive or submunition components, we have to account for the difference in mass, leading to Table 7. This compares impact kinetic energy and the TNT equivalence (in simple energy terms) for the various options. The larger diameter EMRG, with their larger mass, are more desirable, although it is important to note that even the most significant impact is only equivalent to 2.38kg of TNT.

It should be noted, however, that TNT-equivalence is a simplistic model of lethality. The higher impact velocities of smaller projectiles may lend themselves to the use of kinetic submunitions, as they are in the same range of velocity as the GAU-8A cannon, capable of defeating armoured vehicles.

Table 7: Impact energy and TNT equivalence for low-drag projectile options

		REDUCED DRAG MODEL					
		Heavy1	Heavy2	Med HV	Med LV	Light	UAV
Dia	mm	127	97	70	70	57	127
MV	m/s	2000	2000	2000	1500	1500	1000
Pr Mass	kg	18.18	12.73	8	8	3.9	10
	Elevation						
KE impact	10	2.38	2.06	1.61	1.50	0.53	0.66
(MJ)	20	2.33	2.06	1.65	1.30	0.45	0.48
	30	3.42	3.19	2.67	1.70	0.57	0.53
	40	5.22	4.86	3.96	2.35	0.78	0.62
	50	6.99	6.31	4.97	2.94	1.01	0.72
	60	8.39	7.40	5.70	3.36	1.18	0.82
	70	9.36	8.16	6.18	3.63	1.31	0.89
	80	9.96	8.58	6.46	3.79	1.38	0.94
kg TNT	10	0.57	0.49	0.38	0.36	0.13	0.16
(kg)	20	0.56	0.49	0.40	0.31	0.11	0.12
	30	0.82	0.76	0.64	0.41	0.14	0.13
	40	1.25	1.16	0.95	0.56	0.19	0.15
	50	1.67	1.51	1.19	0.70	0.24	0.17
	60	2.01	1.77	1.36	0.80	0.28	0.20
	70	2.24	1.95	1.48	0.87	0.31	0.21
	80	2.38	2.05	1.55	0.91	0.33	0.23

HYPNOS also generates plots of trajectories and velocities, as shown in Figure 6 for the 57mm option. This illustrates the significant reduction in velocity – from 1500 to ≈ 800m/s – over the 20km self-defence range envelope. However, it should be noted that this may be a similar velocity to "boost-coast" type self-defence missiles.

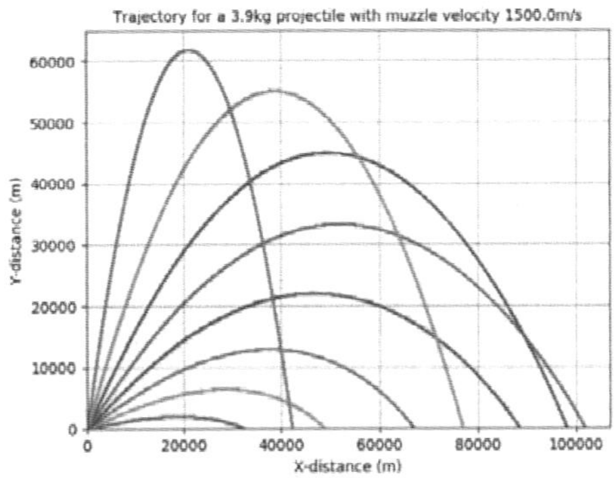

Figure 6a: 57mm EMRG trajectories

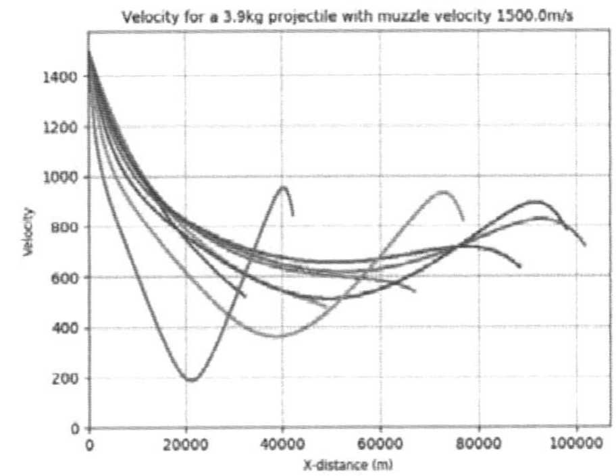

Figure 6b: 57mm EMRG velocities evaluated by HYPNOS

5. THE DESIGN TOOL: ZEOLIT

5.1 BACKGROUND TO THE TOOL

The UCL MRG has an established research theme of concept design methodology for complex ships, most notably the Design Building Block Approach (DBBA), an architecturally-centred approach to early stage design first outlined by Andrews [16], subsequently implemented within the Paramarine software and recently in a Javascript web application [17]. Design investigations using the DBBA involve the development of a small number of "point" studies, utilising the architectural approach to integrate a wider range of ship design aspects beyond those available to a purely numerical approach. There are, however, approaches to early stage ship design which combine architectural aspects and wide ranging numerical exploration, including the TUDelft Packing Approach [18] and University of Michigan developed Intelligent Ship Arrangements (ISA) [19], albeit with limitations on detail, or significant computational overheads.

A new UCL approach has been implemented as an Excel-based tool known as "ZEOLIT" (Zonal Exploration Of Layout and Impacts Tool). This follows on previous, far more sophisticated PhD level UCL work on a library-based approach to concept ship design, in which a wide range of possible design solutions were generated by combining pre-calculated "partial" solutions from a library [20]. Examples of such partial solutions included a range of hullform types varying in displacement and proportions; or different machinery topologies. The current state of the concept design would thus be represented by a "cloud" of possible design points, which could be refined over time. This approach has conceptual similarities to the "Set Based Design" method [21].

5.2 REPRESENTING LAYOUTS

An objective in the development of ZEOLIT was to include layout issues in a fast and wide ranging design exploration. The ship's layout is represented as a grid of 32 layout zones; 8 divisions along the length of the ship and four vertically (upper superstructure, No 1 deck, No 2 deck and lower hull). Spaces and equipment (equivalent to Design Building Blocks, DBB) are assigned to these layout zones, as described in Table 8.

Table 8: Layout modelling in ZEOLIT

Reference Volume

	Stern H	G	F	E	D	C	B	Bow A	19393
2 SS H	799	992	1338	922	1370	816	749	309	7295
1 SS L	420	522	704	485	721	430	394	163	3840
-2 No 2	446	554	747	515	764	456	418	172	4072
-3 H L	123	397	728	712	977	734	432	83	4186
19393	1788	2465	3517	2635	3832	2435	1994	727	

Pre-calculated hull option, showing the possible internal volume for the hull and a notional full-height-full-width single block superstructure.

Geometry Disposition

	Stern H	G	F	E	D	C	B	Bow A	20
2 SS H	0	0	0	0	0.5	0	0	0	0.5
1 SS L	0	0	1	1	1	0.5	0	0	3.5
-2 No 2	1	1	1	1	1	1	1	1	8
-3 H L	1	1	1	1	1	1	1	1	8
20	2	2	3	3	3.5	2.5	2	2	
Number	21								

"Mask" of the available envelope, which defines how much of the pre-calculated hull is actually used. In this example the hull has a single superstructure block amidships with a small bridge deckhouse (layout zone D-2).

Item Distribution Inspector

4

	Stern							Bow	
	H	G	F	E	D	C	B	A	1
2 SS H	0	0	0	0	0	0	0	0	0
1 SS L	0	0	0	0	0	0	0	0	0
-2 No 2	0	0	0	0	0	0	0	0	0
-3 H L	0	0.25	0.1	0	0.25	0.4	0	0	1
1	0	0.25	0.1	0	0.25	0.4	0	0	

Inspecting Liquid Fuel Tanks

Number 4

Spaces and equipment are then assigned to layout-zones. Distributed items, such as fuel tanks, can be assigned to multiple zones as fractions. Alternatively, a pseudo-random assignment could be used where the designer specifies the probability of being in a certain layout zone.

Current Volume

FG: move

	Stern							Bow	
	H	G	F	E	D	C	B	A	2406
2 SS H	0	0	0	0	187	0	0	0	187
1 SS L	0	0	0	0	0	0	0	0	0
-2 No 2	0	0	104	27	0	0	0	0	131
-3 H L	255	187	355	759	258	274	0	0	2088
2406	255	187	459	786	445	274	0	0	

The current volume required in each layout-zone can be calculated, shown here for the entire "move" function (i.e. machinery, fuel tanks etc.)

Current Total Volume

Update totals

	Stern							Bow	
	H	G	F	E	D	C	B	A	9280
2 SS H	0	0	0	0	332	0	0	0	332
1 SS L	0	9	575	447	474	163	9	0	1676
-2 No 2	448	496	621	428	647	401	273	443	3758
-3 H L	275	435	504	954	381	466	303	196	3514
9280	723	940	1699	1830	1834	1030	584	639	

This aggregation can be repeated over all functions and all layout zones to calculate the total volume (and weight) in each zone.

5.3 ZEOLIT CALCULATIONS

The hull reference library currently included in ZEOLIT contains 5832 monohull forms, between 500 and 4000 tonnes, with variations in various hullform parameters. The library contains dimensions, volumes, basic hydrostatics and a simplified power-speed curve between 3.5 and 35 knots. The numerical approach used in ZEOLIT is straightforward; the specified layout and sizing equations for each DBB are applied to each hull in the library in turn. If the balance of overall volume required and available, and weight and displacement, fall within user-specified limits, then the results are retained. Figure 7 illustrates a typical plot of values for the overall volume and mass ratios, illustrating that ZEOLIT can provide a range of options even with a small window of acceptability (+/-5% in this case).

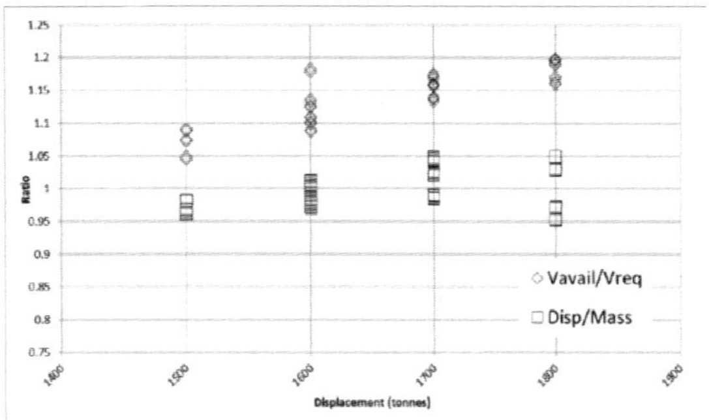

Figure 7: ZEOLIT output, showing the variation in volume and mass ratios within the defined range.

ZEOLIT can make use of any type of model that can be implemented in Excel or via VBA. The ATK-R studies used a mix of; single-line sizing algorithms; historically-derived scaling relationships; single-sheet engineering sizing models (e.g. ESS) and look-up tables to select from a range of precalculated options (e.g. GT and DG selection). The resulting values of volume and weight for each DBB are defined by three point estimates (low, typical, high), and calculations can be run with each of these values or a random mix. The random mix can also be run multiple times to generate a dense cloud of possible solutions. The objective being to assess the impact of design uncertainty in a more useful manner than conventional weight and space margins, which can become "self-fulfilling".

It should be noted that ZEOLIT does not attempt to iterate the design to numerical balance, or make any layout decisions (other than those that can be coded in terms of simple *IF…THEN* statements in Excel). It simply applies the current layout to every possible hull. Although ZEOLIT will not produce the variety and detail of layouts that emerge from an automated approach such as ISA or the Packing Approach, such numerical methods require extensive and complete pre-definition of arrangement rulesets, computation times that do not always align with very early stage design exploration, and can produce layouts of questionable quality [22].

5.4 ANALYSING RESULTS IN ZEOLIT

Each hullform option has a unique ID, permitting a design option to be inspected, but the ZEOLIT approach focusses on comparing the "cloud" of design options, as shown in Figure 7, to understand overall trends. These can include stability and hullform parameters, such as the GMt(solid) calculated by the tool, or the frequency with which each hullform parameter appeared in the acceptable results. Table 9 illustrates the latter; each input parameter occurs an equal number of times in the database, but it can be seen that the Circular M = 8, B/T = 3.5 and T/D = 0.5 are more likely to be acceptable.

Table 9: Analysis of hullform parameter occurrence in the set of acceptable hullforms

Input Values			
Circular M	B/T	T/D	flare
6	3	0.4	6
7	3.5	0.45	
8	4	0.5	8
Occurance % Counter			
0	33	0	50
33	50	0	
67	17	100	50

ZEOLIT records the ratio of available to required volume for each layout zone. This is assigned one of four user-defined levels; *much too small (<80%), slightly too small (80-95%), satisfactory (95-110%), slightly too big (110-120%)* and *far too big (>120%)*. The percentage of possible design options in these ranges is aggregated and displayed as shown in Figure 8.

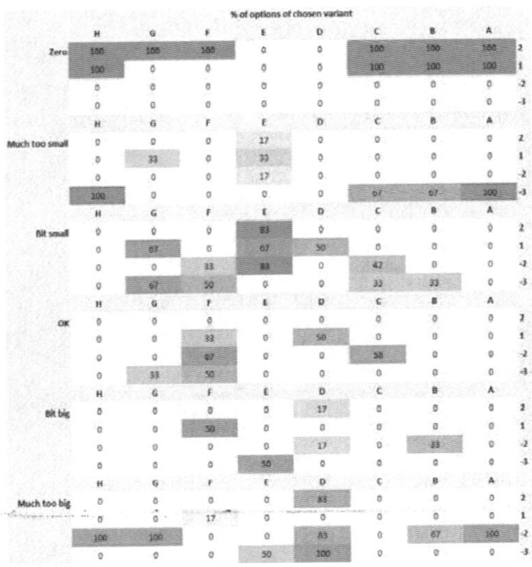

Figure 8: Visualisation of the percentage of options meeting volume criteria for each layout zone

6. POWER AND PROPULSION SYSTEMS (P&PS)

6.1 FUTURE P&PS TECHNOLOGIES

Electrical power generation on warships is currently commonly provided by mechanical prime movers driving synchronous generators providing fixed frequency AC voltage. Energy density of electrical machines is improving over current induction motors through the use of advanced technologies including; water-cooled wound field; permanent magnet; and high temperature superconducting machines. Another approach is the use of high-speed generators, matched to the prime mover speeds, with frequencies of 200-400Hz.

For both electrical machines, running at higher speeds results in higher power density than conventional 60Hz counterparts [23]. Although high speed motors will require a gearbox to drive propellers, the reduction in motor size and weight can be sufficient to offset the need for the gearbox [24].

Fuel cells are an alternative power source for future warships [25]. Proton Exchange Membrane Fuel Cells (PEMFCs) are being used in submarine applications, but require high purity hydrogen, necessitating reformation from hydrocarbon fuels. However overall efficiency ranges from 50-60%. Solid Oxide Fuel Cells (SOFCs) however operate at high temperatures (500-1000°C) and can use hydrocarbons via internal steam reformation to generate hydrogen. Total fuel efficiency of SOFCs can reach 85% when integrated with heat recovery systems enabled by the high operating temperature [26]. However, high temperature operation incurs long start up times and low ramp rates, requiring an alternative fast-acting power source to handle transient demands. SOFC suitability for hydrocarbon fuels may be significant if synthetic hydrocarbons produced using renewable energy sources (so called "electrofuels") such as methanol become a significant part of the marine fuel landscape.

The use of DC distribution architectures for naval vessels is attractive for integration of pulsed loads such as EMRGs, owing to the lack of requirement for synchronisation between power sources as in AC systems and subsequent resistance to transient loads. Contemporary warship power distribution systems are primarily based upon 3-wire AC distribution [27] and the AC-DC converters required for each rotating machine would incur increased losses and cooling demands. The use of Silicon Carbide semiconductor devices could reduce switching and conduction losses [28] but at higher unit cost. DC distribution can still suffer power system instability when high ramp loads induce voltage drop caused by unbalanced power flow, with advanced energy management being a possible solution [29].

6.2 ATK-R SYSTEM TOPOLOGIES

Table 10 summarises the topologies examined so far using ZEOLIT. All propulsion motors are high-speed geared machines, with generators being current off-the-shelf systems for DG and high-speed machines for GTA. Where the speed is marked with "+", the diesels / DGs are sized for that speed, with the GTA being fixed in size and maximum speed thus being "whatever you get" from the combined total power.

Table 10: ATK-R P&PS system topologies

Topology	*Max Speed*	*Type*	*Power distro*	*Propulsion*	*Prime movers*	*ESS*
1a	35	Hybrid	DC	2 x propellers driven by high speed geared motors 1 x waterjet	2 x small DG 2 x medium DG 1x high speed GTA	Hybrid battery/ supercapacitor
1b	35	-ditto-	-ditto-	2 x propellers driven by high speed geared motors 1 or 2 x waterjet	2 x small DG 2 x medium DG 1 or 2 x high speed GTA	-ditto-
1c	25+	-ditto-	-ditto-	-ditto-	2 x small DG 2 x medium DG 2 x high speed GTA	-ditto-
1d	-ditto-	-ditto-	-ditto-	4 x propellers driven by high speed geared motors	-ditto-	-ditto-
2a	25	IFEP	-ditto-	2 x propellers driven by high speed geared motors	2 x small DG 2 x large DG	-ditto-
2b	25	-ditto-	-ditto-	- ditto -	2 x small SOFC 2 x large SOFC	-ditto-
2c	25	-ditto-	-ditto-	-ditto- plus 1 or 2 x waterjet	-ditto- plus 1 or 2 high speed GTA	-ditto-
3	25+	Mech	-ditto-	2 x propellers driven by high speed diesels 2 x waterjets driven by GT	2 x small DG (hotel) 2 x large diesel 2 x high speed GTA	-ditto-

6.3 TOPOLOGIES OF INTEREST

6.3(a) Topology 1a,b,c

Topologies 1a, b and c are CODLAG systems, with the example of 1a shown in Figure 9. When operating the EMRG, the waterjet is declutched and the GT is connected to a high frequency generator set via power take-off (PTO) to charge the supercapacitor based EMRG ESS. A separate battery-based ESS is sized to support charging of the EMRG, depending on the GT size and scenario specifics.

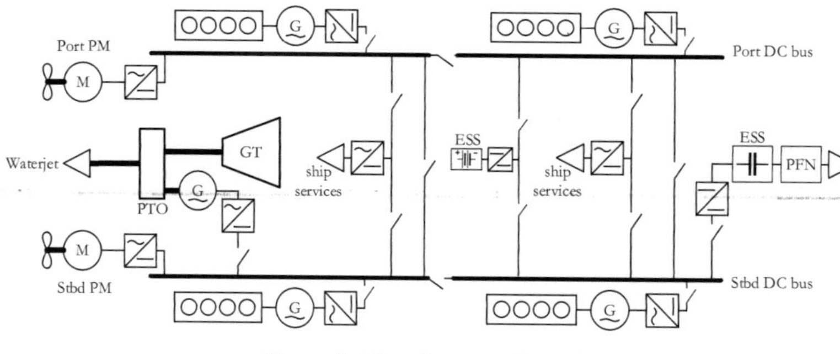

Figure 9: Line diagram of topology 1a

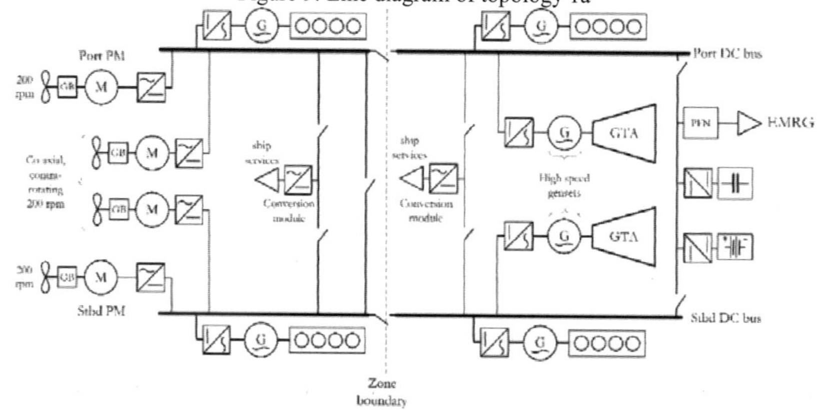

Figure 10: Line diagram of topology 1d

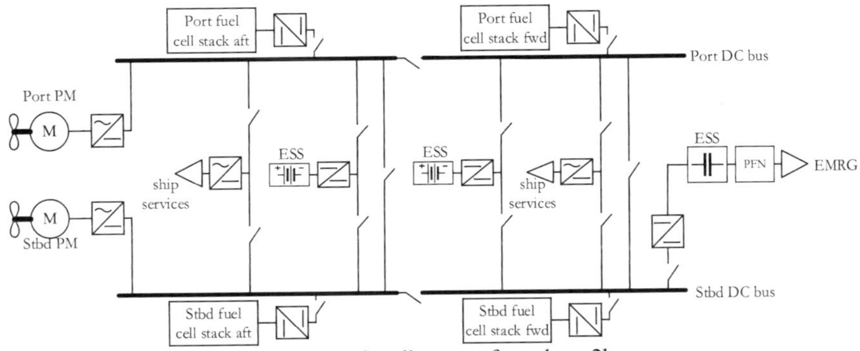

Figure 11: Line diagram of topology 2b

6.3(b) Topology 1d

Topology 1d is an IFEP system, developed in response to smaller ATK-R variants having insufficient hull volume aft to accommodate boost waterjets, but insufficient hull draft or internal volume to easily accommodate high power propellers or propulsion motors. This option is illustrated in Figure 10. The High-speed GTAs are connected to a DC IFEP distribution system, and the boost waterjet is replaced by two geared motors driving a centreline co-axial contra-rotating propeller arrangement.

Contra-rotating propellers were selected to allow a greater propeller area to be accommodated in the small cross-sectional area available on smaller ATK-R variants. The use three- and four-shaft arrangements on smaller fast warships being an established practice, e.g. Danish Flyvefisken-class patrol vessels. Contra-rotating propellers have been claimed to offer efficiency improvements [30], but for this study they were assumed equal to conventional propellers of the same diameter, speed and loading, as this issue is complicated by detail design and hullform interaction.

6.3(c) Topology 2a,b

Topologies 2a and b dispense with the high power GT and rely on four DGs or methanol-fuelled SOFC stacks. The lower installed power necessitates a substantial battery ESS to support EMRG charging, as shown in Figure 11. For this study, the SOFC stacks were assumed to be unitised in modules similar to DG sizes. SOFC were assumed to have an efficiency of 60%, as they are still at an early stage of commercial application [31], although highly detailed studies have been performed on some applications, e.g. aircraft APUs [32].

6.4 MODELLING ESS IN ATK-R

The characteristics of Energy Storage Systems (ESS) are covered in more detail in Pawling et al [6] and the use of time domain simulation to model them in [7]. ESS are crucial to the integration of high-energy weapons and their associated pulsed loads for two reasons; firstly they buffer the load seen by the wider P&PS thus mitigate adverse QPS conditions [4] and secondly they allow a vessel with lower installed generating power to support a high power load.

Two types of ESS were used; a dedicated "weapon" system, utilising supercapacitors, and a potentially multi-purpose battery ESS. The supercapacitors were sized using data from Skeleton Technologies [33] and the energy stored by the complete unit was determined by the scenarios provided in Table 4. Specifically the "burst" scenarios implied that the EMRG fired fast enough that the dedicated ESS could not be recharged between shots, whereas the "continuous" scenarios involved sufficiently low RPM to permit recharging of the weapon ESS. The supercapacitor ESS model is described in Pawling et al [6] and included considerations such as minimising inrush current and voltage limits per string.

The battery based ESS was sized using data from Plan B Energy Storage [34], using Lithium-Ion NMC cells. The ESS sizing model generates options based on required power and energy, with ZEOLIT selecting the largest and applying a weight and space margin for converters, enclosures etc. The battery-based ESS has alternative uses; in the SOFC options it could be used to handle non-weapon ramp loads. In IFEP configurations, it may be store enough energy to allow for an operationally useful "low signature" mode. The ATK-R studies reported here aimed to scope the likely stored energy; if such a general-purpose ESS were to be adopted, some segregation of the ESS into smaller units may be required firstly to manage battery performance and secondly to meet the current and voltage limits necessitated by the battery system and converter interface.

7. ATK-R SPECIFICATIONS

The ATK-R project is focussed on two main options; a "large" vessel, with complement and endurance sufficient to support overseas deployment, and a "small" vessel, with reduced combat system and limited endurance. The large options were influenced by OPVs such as the RNZN *Protector* class OPV and Russian *Project 2038.5/6* corvettes, whilst the small option is a railgun equivalent of the Russian *Project 21631;* a small patrol vessel with strategic implications due to its carriage of a long range potentially nuclear-capable missile system. The basic specifications for these options are outlined in Table 11.

The different combinations of machinery and weapons options investigated so far are summarised in Table 12 below. This paper will focus on those vessels equipped with a single "Heavy 1" type EMRG.

Table 11: Common specifications for ATK-R

Complement	Large: 50 Small: 39	*Deck height*	2.75m
Stores endurance	Large: 30 days Small: 10 days	*DC zones*	3
Fuel endurance at 14 knots	Large: 5000nm Small: 2300nm	*Boats and craft*	1 x 7m RIB Large only: UxV bay for UAV

Command	*Radar*	*ASW*	*AAW*	*EW*
15-seat ops room	Large: 4-face S-band MFR in integrated mast Small: Rotating S-band MFR	Hull mounted sonar	24 x CAMM (or EMRG)	ESM system
Small warship comms fit	*Electro-Optical* 2x trackers	24 x 210mm rocket-torpedo	*Small Calibre Guns* 2 x 30mm, 5000 rnds	ECM system Active decoys, 80 rnds

Table 12: Machinery and ship specification combinations

	1 x Heavy 1 EMRG		2 x Medium HV EMRG
MCY Option	Small	Large	Large
1a/b	-	X	X
1c	X	X	-
1d	X	X	-
2a	X	-	-
2b	-	X	-
3	X	-	-

8. ATK-R DESIGN STUDIES

8.1 LARGE DESIGNS

Table 13 compares the resulting displacements of the "large" ATK-R options, with a single "heavy 1" EMRG and a battery ESS storing energy for a single engagement. Each has been assigned an ID letter for this discussion. By necessity the discussion here can only cover the key points, rather than every option in depth.

Table 13: Comparison of large design options, single Heavy 1 EMRG, single-engagement ESS

	ID	**A**	**B**	**C**	**D**	**E**	**F**	**G**	**H**	**I**
	Machinery	**1a/b**	**1a/b**	**1a/b**	**1a/b**	**1d**	**1d**	**2b**	**2b**	**2c**
	GT	Various	2 x MT-7	1 x LM2500	1 x LM2500	2 x MT-7	*2 x MT-7*	None	2 x MT-7	2 x MT-7
	Firing speed, knots	21	24	21	24	24	*24*	14	21	24
	Va/Vr limits	0.95-1.05	0.95-1.05	0.95-1.05	0.95-1.05	0.95-1.05	*0.9-1.1*	0.95-1.05	0.95-1.05	0.95-1.05
	Disp/Mass limits	0.95-1.05	0.95-1.05	0.95-1.05	0.95-1.05	0.95-1.05	*0.9-1.1*	0.95-1.05	0.95-1.05	0.95-1.05
	# Options	36	12	54	20	2	*44*	34	36	25
Displacement, te	**Max**	2800	1900	2000	2200	2000	*2200*	2000	2000	2200
	Min	1900	1700	1700	1800	2000	*1800*	1600	1600	1700
	Mean	2328	1817	1804	1970	2000	*1991*	1732	1756	1888
	Median	2250	1800	1800	2000	2000	*2000*	1700	1700	1900
	Mode	2000	1800	1800	2000	2000	*1900*	1700	1700	1900

8.1(a) Wide-Ranging Survey: Option A

For option A, the complete capability of ZEOLIT was used to select boost GT options as well as cruise DG.

The wide range of incremental machinery choices, when combined with the parametric ESS sizing, lead to a broad and non-uniform set of options within the limits of numerical balance, as illustrated by Figure 12. Two groups of solutions appear, with no viable options in the 2400-2500 tonne range, due to a lack of suitable GTs. It is important to compare this spread of results with the statistics in Table 13; the larger designs are actually outliers, and the single most common displacement is 2000 tonnes.

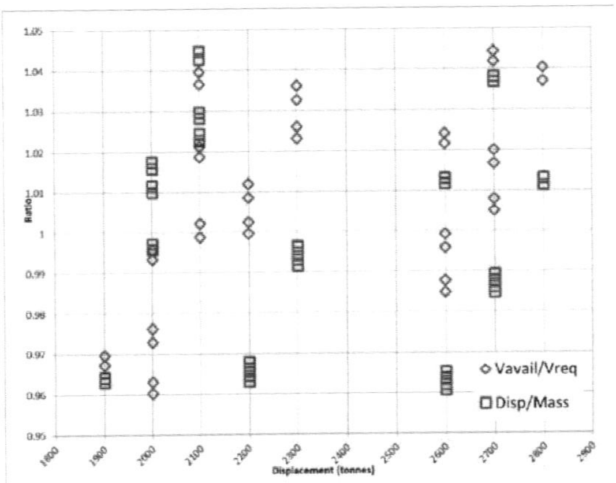

Figure 12: Variation in volume and mass ratios for option A.

The installed power varies accordingly, and above 2100 tonnes displacement the GTs are powerful enough to charge the EMRG "weapon" ESS without battery boost. Below 2100 tonnes, a 7MW, 0.78MWhr battery with a mass of 34 tonnes is required, which can be accommodated forward in the lower hull. Figure 13 shows the variation in DG and GT power with size.

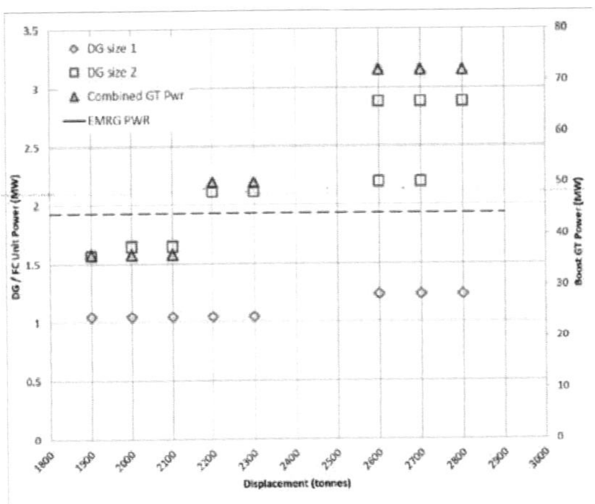

Figure 13: Variation in machinery unit power for option A.

Fixing the GT size reduces the displacement, with the smallest GT-equipped option having a displacement of 1700 tonnes. Reducing the speed to be maintained whilst firing from 24 to 21 knots permits the ESS to be reduced in size and the ship overall by 200 tonnes. Considering the LM2500 options, however, a key issue emerges; although it is possible to find sufficient volume in the hull amidships for the GTA (as shown in Figure 14), considering the modal waterline length is only 98.3m, accommodating a 14m long GTA and meeting combatant subdivision requirements will be difficult, so the option of two smaller GT (MT-7) was adopted.

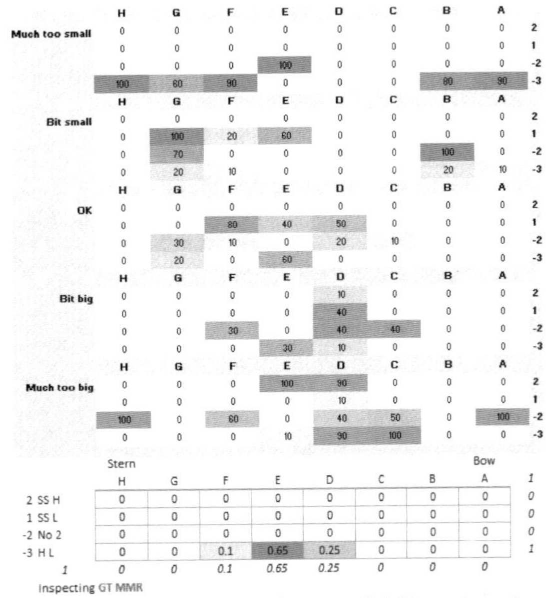

Figure 14: Aggregated volume availability ratio and distribution of GT machinery room in option A

8.1(b) Fuel Cells and Methanol: Options G, H and I

Comparing options B and I, we see that a methanol-fuelled fuel-cell option may add as little as 100 or as much as 300 tonnes to the overall displacement of the vessel, with the former being more likely, making a liquid electro fuel a viable possibility from this very high-level view. However, this may be due to the choice of endurance range and operating profile for ATK-R, so further studies would be required to determine if this holds true over a range of design assumptions.

8.1(c) Battery ESS Sizes

Choosing a smaller GT leads to a much larger battery ESS to support EMRG charging. Option I, for example, requires a 33MW, 3.64MWhr battery system with a mass of 53.2 tonnes. Although this would store enough energy to permit the vessel to operate at 10 knots for up to 3 hours, it is very difficult to accommodate in the forward hull near the EMRG, as shown in Figure 15.

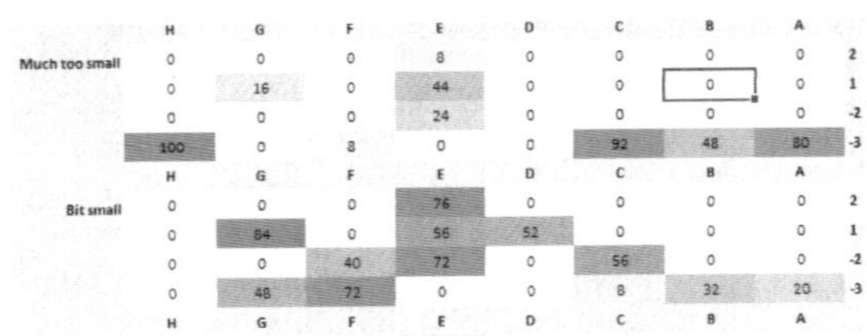

Figure 15: Visualisation of layout zones with insufficient internal volume for option I

This layout issue effectively restricts the ESS to being sized for a single engagement only. This will have an impact on the CONOPS for the vessel as even a high-discharge battery design can only be charged much more slowly. However, this limitation on recharge rate also means that substantial GTA power may be available for repositioning, in a "shoot-and-scoot" mode of operations.

A layout advantage that the battery ESS does have is that it can be divided into several units, in different watertight compartments, thus avoiding the length issue of large GTAs. This has implications for volume efficiency and cost (due to duplication of enclosures, switchboards etc.) but may be desirable if the battery ESS can be made a "general purpose" system providing, for example, reversionary zonal power supplies.

8.2 SMALL DESIGNS

With a slower maximum speed requirement of 24 knots (the GT boost speed being regarded as a "bonus"), the smaller designs did not make use of the GT selection mechanism. Instead, where GT were used, they were fixed to a particular type. Most design options required an increase in permissible volume and mass imbalance, from +/-5% to +/-10% to generate a useful range of options. Table 14 summarises the options for the "small" ATK-R options, with a single "heavy 1" EMRG and a battery ESS storing energy for a single engagement.

The first point of note is that the smallest option, N, uses geared diesel drive. It is approximately 300 tonnes heavier than the Russian *Project 21631* corvette that is its inspiration. Of this, the EMRG, battery and twin MT-7 GT make up the main part, at 193 tonnes, with the battery being 53.2 tonnes.

Table 14: Comparison of small design options, single Heavy 1 EMRG, single-engagement ESS

	ID	**J**		**K**		**L**		**M**		**N**	
	Machinery	**1c**	**1c**	**1c**	**1c**	**1d**	**1d**	**2a**	**2a**	**3**	**3**
	GT	2 x MT-7	*2 x MT-7*	1 x LM2500	*1 x LM2500*	2 x MT-7	*2 x MT-7*	None	*None*	2 x MT-7	*2 x MT-7*
	Firing speed, knots	24	*24*	24	*24*	24	*24*	24	*42*	24	*24*
	Va/Vr limits	0.95-1.05	*0.9-1.1*	0.95-1.05	*0.9-1.1*	0.95-1.05	*0.9-1.1*	0.95-1.05	*0.9-1.1*	0.95-1.05	*0.9-1.1*
	Disp/Mass limits	0.95-1.05	*0.9-1.1*	0.95-1.05	*0.9-1.1*	0.95-1.05	*0.9-1.1*	0.95-1.05	*0.9-1.1*	0.95-1.05	*0.9-1.1*
	# Options	2	*24*	0	*10*	4	*47*	12	*24*	18	*81*
Displacement, te	**Max**	1500	*1600*	-	*2200*	1600	*1800*	1400	*1500*	1400	*1500*
	Min	1500	*1400*	-	*2100*	1500	*1400*	1300	*1200*	1300	*1200*
	Mean	1500	*1500*	-	*2160*	1550	*1602*	1350	*1338*	1333	*1332*
	Median	1500	*1500*	-	*2200*	1550	*1600*	1350	*1300*	1300	*1300*
	Mode	1500	*1500*	-	*2200*	1500	*1500*	1300	*1300*	1300	*1300*

Comparing option N with J and L, (IFEP options), the change from geared to electric drive increases the modal displacement by 200 tonnes, due to additional machinery and the associated volume. Deleting the GT of option N and relying only on a battery ESS with a small (2MW) shaft generator for recharge leads to option M, with no change in displacement as the battery mass increases to 68.4 tonnes. Due to the small size of the boost GT, maximum speeds are around 28 knots.

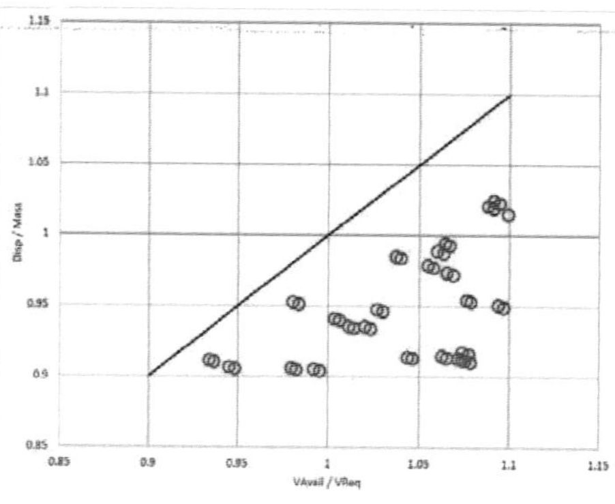

The smaller ATK-R options with electrical propulsion are weight dominated; options will generally have excessive overall volume, as illustrated in Figure 16. This results from the hullform displacement being driven by the mass of the electrical machinery, whilst the upper hull and superstructure are occupied by spaces such as accommodation, constant across the variants. Figure 17 illustrates how even with excess volume overall; the designs can still have local issues with the arrangement.

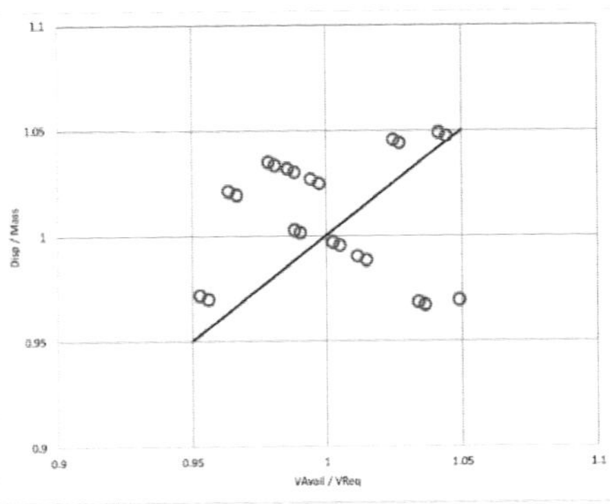

Figure 16: Comparison of Volume and Mass ratios for small option L (IFEP, upper) and N (mechanical, lower), with unity slope indicated

	H	G	F	E	D	C	B	A	
Much too small	0	0	0	0	6	0	0	0	2
	0	4	0	0	0	0	0	47	1
	17	4	0	0	0	0	0	0	-2
	60	0	4	21	0	0	0	0	-3
	H	G	F	E	D	C	B	A	
Bit small	0	0	0	49	77	0	0	0	2
	0	77	0	13	79	0	0	53	1
	13	43	53	40	40	77	0	68	-2
	6	0	47	47	4	0	0	21	-3
	H	G	F	E	D	C	B	A	
OK	0	0	0	47	17	0	0	0	2
	0	19	34	68	21	21	87	0	1
	26	53	47	51	53	23	15	32	-2
	0	9	30	32	4	0	0	9	-3
	H	G	F	E	D	C	B	A	
Bit big	0	0	0	4	0	0	0	0	2
	0	0	45	19	0	51	13	0	1
	0	0	0	9	6	0	64	0	-2
	0	9	19	0	17	0	0	4	-3
	H	G	F	E	D	C	B	A	
Much too big	0	0	0	0	0	0	0	0	2
	0	0	21	0	0	28	0	0	1
	45	0	0	0	0	0	21	0	-2
	34	83	0	0	74	100	100	66	-3

Figure 17: Aggregated volume availability ratio for design variant L

9. SUMMARY AND CONCLUSIONS

The Advanced Technology Korvette – Railgun (ATK-R) is an ongoing project at UCL; a thought experiment concerning future small warships armed with electric weapons. It brings together three main areas of research; design methodology (ZEOLIT); ballistics modelling (HYPNOS); and PhD level research into Energy Storage Systems (ESS). This paper has outlined these three areas of research and provided some examples of the ATK-R design options developed so far.

ZEOLIT and HYPNOS have been demonstrated by application to the ATK-R concept, and the latter has been validated by comparison with third-party calculations. ZEOLIT has demonstrated utility in enabling the generation of a range of concepts, combining parametric numerical models with sufficient layout fidelity to identify some key drivers and issues.

The studies so far indicate that a small, corvette-sized vessel would be capable of supporting an EMRG similar to some of the largest currently under development. The weapon itself is less influential on ship sizing and layout than is the ESS and P&PS required to provide sufficient power and energy for useful engagement scenarios. Although these increase the size of the vessel compared to conventionally armed ships, battery ESS will have sufficient stored energy to support a stealthy "zero emissions" operating mode for a limited period, or provide survivable distributed power. If additional generating capacity is to be installed, pairs of small GTA are desirable due to constraints on layout and subdivision. ATK-R options made use of geared high-speed motors and high-speed generators, and considering that many options still had crowded machinery spaces, this technology, along with ESS, may be vital to integrating electric weapons on smaller ships.

With regards to combat systems, projectile design was found to be the crucial factor, with both impact velocity and range being highly sensitive to projectile drag coefficient. This was particularly the case in the hypersonic regime soon after launch, where much of the launch kinetic energy was lost. If kinetic energy is to be the primary damage mechanism, then specialised low-drag projectiles, with geometries unlike those of current guided and course-corrected weapons, will be required, and this will have implications for aspects such as guidance systems.

10. FUTURE WORK

There are several areas of future work and investigation that have been identified. ZEOLIT is being applied and evaluated by a UCL MSc student for their individual project and this assessment is largely independent from the original developer. Longer-term it is expected that certain aspects of the approach, such as the numerical assessment of uncertainty in preference to simple design margins, may be adopted in UCL design and teaching activities, whilst the prototype Excel ZEOLIT implementation may be superseded by dedicated software.

HYPNOS currently only models ballistic unguided projectiles, and future developments will add large-yaw angle analysis to permit the modelling of terminally guided projectiles. Assessment of a wider range of drag models and projectile geometries will not only improve the model, but feed back into EMRG weapon sizing, to replace current assumptions which have their origins in existing guided projectiles. As with ZEOLIT, the combat systems component has a link to an MSc individual project, using the UCL Rapid Vulnerability Assessment Model (RVAM) to explore and characterise lethality of future electric weapons.

The ATK-R study has reinforced the importance of Energy Storage Systems and P&PS integration in the electrification of naval weapons. Further characterisation of the physical and configurational options for ESS is required, including an understanding of the implications of splitting a high-power battery ESS or using advanced technologies with higher discharge rates.

11. ACKNOWLEDGEMENTS

Development of ZEOLIT was carried out as part of the Preliminary Ship Design General Arrangements NICOP project and its successor NICOP, both funded by Ms. Kelly Cooper from the US Navy Office of Naval Research, and this support is gratefully acknowledged. The conclusions presented are the opinions of the authors and do not represent MoD or DoD policy.

12. REFERENCES

1. Simmonds O.J.: "Advanced hybrid systems and new integration challenges", in International Naval Engineering Conference,2016, Bristol: IMarEST, pp. 1–10.
2. Andrews D.J. Bucknall R. & Pawling R.J.: "The Impact of Integrated Electric Weapons on Future Warship Design", INEC 2010, IMarEST, Portsmouth, UK.
3. Whitelegg I.: "Power system design considerations when integrating electromagnetic railguns with electric warships", In: INEC 2016, Bristol: IMarEST.
4. Whitelegg I.: "On the integration of electromagnetic railguns with warship electric power systems", PhD Thesis, UCL, 2016.
5. Andrews D.J. Bucknall R. Pawling R.J. Greig A. & McDonald T.: "The Impact of Integrated Electric Weapons on Future Warship Design Using Conventional and Unconventional Hullforms", Engine as a Weapon (EAAW) 2011, IMarEST, London.
6. Pawling, R.J., Farrier, L. & Bucknall, R.: "The Advanced Technology Corvette - Railgun (ATK-R) Design Study - Future Weapons and Small Ship Power Systems", In: International Naval Engineering Conference 2018 (accepted). Glasgow: IMarEST
7. L. Farrier. 2018., Investigating the faulted performance of warship power systems with integrated energy storage. In: International Naval Engineering Conference 2018 (accepted). Glasgow: IMarEST
8. Office of Naval Research (ONR): "Naval EM Railgun Innovative Naval Prototype", Precision Strike Armaments Precision Strike Armaments Technology Fire Power Forum Technology Fire Power Forum, June 2008.
9. General Atomics: "Missile Defense Systems – Electromagnetic Gun Systems", http://www.ga.com/missile-defense-systems, 2018.
10. "Rethinking the Naval Gun System", OTO Melara brochure
11. "Maritime unmanned air vehicle in production", RINA Warship Technology, March 1998, p. 16
12. NASA. 2015. Earth Atmosphere Model – Metric Units. [online] Available at <https://www.grc.nasa.gov/WWW/K-12/airplane/atmosmet.html>
13. McCoy, R. L., 2012. Modern Exterior Ballistics. 2nd ed. Altglen PA: Schiffer.
14. Walls, W. A., Weldon, W. A., Pratap, S. B., Palmer, M. and Adams, D., 1999. Application of electromagnetic guns to future naval platforms. IEEE Transactions on Magnetics, vol. 35, no. 1, pp. 262-267. 10.1109/20.738414
15. McNab, I.R., Fish, S. & Stefani, F,, "Parameters for an Electromagnetic Naval Railgun", IEEE Transactions On Magnetics, Vol. 37, No. 1, January 2001
16. Andrews D.J.: "A Comprehensive Methodology for the Design of Ships (and Other Complex Systems)", Proceedings of the Royal Society, Series A (1998) 454, p.187-211, Jan 1998.
17. Pawling R.J. Piperakis A.S. & Andrews D.J.: "Developing Architecturally Oriented Concept Ship Design Tools for Research and Education", International Marine Design Conference (IMDC) 2015, Tokyo, May 2015.
18. Van Oers B.J.: "A Packing Approach for the Early Stage Design of Service Vessels", PhD Thesis, Delft University of Technology, Delft, The Netherlands, 2011.
19. Daniels A.S. Tahmasbi F. & Singer D.J.: "Intelligent ship arrangement (ISA) methodology improvements and capability enhancements", IMDC 2009, Trondheim, Norway, May 2009.
20. McDonald T. Andrews, D.J. & Pawling R.J.: "A Demonstration of an Advanced Library Based Approach to the Initial Design Exploration of Different Hullform Configurations", Computer-Aided Design, Vol. 44, Issue 3, March 2012, pp. 209-223.
21. Singer D.J. Doerry N. & Buckley M.E.: "What is Set-Based Design?", Naval Engineers Journal, Vol 121/4, October 2009.
22. Droste, K., Kana, A.A. & Hopman, H.H., "Process-based Analysis of Arrangement Aspects for Configuration-Driven Ships", IMDC 2018, Helsinki, June 2018
23. Calfo R.M. et al.: "High-speed generators for power-dense, medium-power, gas turbine generator sets", Naval Engineers Journal, 120 (4), 45–54, 2008.
24. Chowaniec Z. Forbes J. & Shaw J.: "High speed motor propulsion for warships", In: International Naval Engineering Conference, 2012, IMarEST.
25. Doerry N. Amy J. & Krolick C.: "History and the Status of Electric Ship Propulsion, Integrated Power Systems, and Future Trends in the U.S. Navy", Proceedings of the IEEE, 103 (12), 2243–2251, 2012.
26. DNV-GL: "Study on the use of Fuel Cells in Shipping", 2017.

27. Meggs R.W. & Pollard C.: "Protective Devices for DC Power Networks", in International Naval Engineering Conference, 2016, Bristol: IMarEST, pp. 1–12.
28. Gattozzi, A.L., Strank, S.M., Pish, S.P., Herbst, J.D., Hebner, R.E., and Engelkemeir, F.D., 2017. Power converter design options for the 12 kVdc bus system. 2017 IEEE Electric Ship Technologies Symposium, ESTS 2017, 246–250.
29. Vu,T.V. Gonsoulin D. Perkins D. Diaz F. Vahedi H. & Edrington C.S.: "Predictive energy management for MVDC all-electric ships", 2017 IEEE Electric Ship Technologies Symposium, ESTS 2017, 327–331.
30. Doerry, McCoy and Martin, 2010.,Energy and the affordable future fleet. In: International Naval Engineering Conference 2010. Portsmouth: IMarEST
31. Kyocera: "KYOCERA Develops Industry's First 3-Kilowatt Solid-Oxide Fuel Cell for Institutional Cogeneration", July 6, 2017, https://global.kyocera.com/news/2017/0702_bnfo.html
32. Whyatt G.A. & Chick L.A.: "Electrical Generation for More-Electric Aircraft using Solid Oxide Fuel Cells", PNNL-21382, US DoE, April 2012.
33. Skeleton Technologies: "Technical specification: Module SMOD170V53F", 2016.
34. Plan B Energy Storage: PBES Specification Sheet, 2017.

13. AUTHOR'S BIOGRAPHIES

Dr. Rachel Pawling is a Lecturer in Ship Design at UCL, teaching the subject of ship design to undergraduate and postgraduate students. She obtained her PhD in computer aided ship design in 2007 and has subsequently continued her research in projects funded by the EC, UK and US governments.

Mr. Luke Farrier is a Marine Engineering Research Assistant at UCL and 2nd year PhD student investigating power system performance when integrating energy storage systems. Prior to this he obtained a degree in Mechanical Engineering from the University of Exeter, and an MSc in Marine Engineering (Electrical) from UCL.

Dr. Nick Bradbeer is a Lecturer in Naval Architecture at UCL, teaching both surface ship and submarine design. He obtained an MSc in Naval Architecture from the department in 2003, then returned to complete his PhD in structural survivability. His research interests include operational modelling and analysis and applications of VR in teaching and design.

VIRTUAL REALITY FOR CONCEPT DESIGN

K. Goh, Knud E Hansen
RJ Spencer, Stirling Labs

SUMMARY

In this paper, the authors investigate the importance of visualisation in the design process and discuss benefits and requirements that can lead to successful outcomes using virtual reality technology for designing warships. The investigation focuses on concept design using Stirling Labs' ShipSpace visualisation and collaboration tool. Results are presented for various tests using ShipSpace to demonstrate that virtual reality is an effective medium for visualising design and forming the basis for a better method for design collaboration.

1. BACKGROUND

The term Virtual Reality or VR is used here to describe a head-mounted computer display (HMD), highly accurate positional tracking system and computer software that updates the display as the user moves with sufficient fidelity that the user can sustain the illusion that they are immersed in the virtual environment displayed. Note that so-called "caves" and 3D theatres, along with the mobile phone devices sometimes referred to as "VR" represent very different technology and do not provide many of the features discussed here.

Historically in the pursuit of design there has been an ongoing drive to find the best way to convey ideas. From the first lines in the sand to ink on papyrus to current day printing. The practice of using drawings with views, elevations and sections is simply an abstraction of the design intended to enable people to understand the design concept so that it can be agreed and constructed.

1.1 PERCEPTION

In the 19th century Helmholtz [1] suggested that we understand the true size of distant objects (that appear small) because the faulty image projected onto our retina is corrected by an unconscious judgement based on facts available to the observer. Arnheim [2] asked if the brain first created a faulty perception based on the raw data that is then interpreted in a manner better fitted to the facts by inferences drawn from the observer's knowledge, or if the given perceptual situation contains aspects that assign the appropriate size to distant objects. Arnheim's theory of Visual Thinking went on to argue that it was the latter, that the interpretation of raw visual data was part of perception itself and not a separate cognitive process that occurred after perception has delivered a deficient message.

1.2 VISUALISATION

Unlike perception, visualisation is a fully cognitive process. Images are created in the mind's eye by consciously thinking about them. Galton [3] showed that there is a wide range of human ability to visualise well known everyday objects. Zeman [4] discussed the case of MX; a UK building inspector who underwent a minor medical procedure and subsequently lost his ability to conjure images in his mind. He had lost the ability to visualise. MX underwent extensive testing, including fMRI scanning to determine if this was a physical incapacity or some psychological phenomena. Zeman concluded that visualisation is indeed a discreet ability and that it can be lost. A further paper [5] coined the term 'Aphantasia' for people that lack or have limited capacity to create mental images.

1.3 IMAGINATION

Interpretation of abstract data such as CAD plans is an additional cognitive skill beyond visualising well known objects or places. Imagination is required to read abstracted drawing information such as elevations, views and sections and mentally process that information as mental images of the object. Shepard & Metzler describe this capability generally as 'Mental Rotation' in [6]. The Mental Rotations Test developed by Vandenberg & Kuse [7] require the subject to interpret various views including flat 2D or 3D drawing representations. These tests correlate strongly with other measures of a person's spatial visualisation capability and have no association with verbal or narrative ability.

Further to the ability to interpret drawings, is a person's ability to imagine things that are not familiar. The authors suggest that people with a high capacity for mental rotation may be better at visualising new designs from drawings, as it appears that this capability is generally more effective than alternative strategies that rely more on memory of similar objects.

McKim [8], another proponent of visual thinking, believes there are three separate kinds of imagery: 1) seeing, 2) imaging and 3) drawing. The interplay between these and the individual's capabilities in each will determine their general capability to visualise concepts.

Mohler [9], in his review of research and technology for improving spatial ability using VR, reasserted Gardner's [10] Theory of Multiple Intelligences and the importance of spatial ability as a form of intelligence. Sorby also notes [11] that there have been numerous educational

research studies into spatial skills that conclude that these learned skills are important to a range of fields, including engineering and design.
The ability to 'read drawings' is a skill that can be learnt but it is clear that the ability of individuals varies hugely, as does their efficiency in this task. The cognitive load required for visualisation can dramatically affect the individual's ability to understand the design and consider inherent issues.

1.4 HETEROGENEOUS SKILLS

During a complex project such as designing a new warship, designers and engineers make up less than half of the professionals involved. Given the above, it would be a significant undertaking to ensure that there are consistent CAD visualisation skills across the entire design & engineering staff. Furthermore, there is no question that the other professions involved in the project will have an even wide variation of capabilities and hence cognitive load in visualising from drawings.

1.5 TRADITIONAL ANALOGUES

a) Artist's Impression & Scale Models

It is very common to employ an artist who is particularly skilled at visualisation to produce 'artistic impressions' – pictures that people can more easily understand. Scale models are also often used, traditionally for informing the exterior design. They have also been used for development of complex spaces such as engine rooms, which need careful design to ensure all equipment and connecting services can be accommodated. Scale models are good at giving an overall impression of shape and form but are expensive and not good at conveying a sense of scale

b) Mock-ups & Prototypes

Full size mock-ups of spaces are used where human interface with the built environment is particularly important. Such areas on a warship include the bridge and a warships operations room or Combat Information Centre (CIC). These can be quite simple with just dumb furniture arranged on a floor layout or quite elaborate with functional work stations and all dimensions of the environment spatially correct. Very similar to a movie set, these mock-ups can be very expensive in terms of space and effort to construct and generally only have limited use and benefits.

1.6 DIGITAL ANALOGUES

With the increasing power of digital computing in the last decades, the ability to create more realistic digital virtual representations of the real world and design has developed tremendously. From using computers as a simple drafting tool for traditional 2D drawings, to the ability to created accurate 3D geometrically accurate forms that can be viewed from any perspective.

At the time of this research, there are dozens of modelling and CAD programs that can create realistic 3D objects and complex virtual environments such as a ship, complete with internal details. Rendering a 3D model to a 2D screen is obviously not VR and requires visualisation beyond mere perception for understanding.

2. LIMITATIONS OF 3D CAD AS A VISUALISATION TOOL

a) Navigation in 3D

The change between designing in two dimensions to designing in three is quite profound. In 2D design, the concept of up, down, left and right is well established and many of our paper and computer-based tools such as your desktop and mouse all are optimised for two-dimensional navigation. Even the computer monitor is a two-dimensional display device.

When working in three dimensions, using traditional tools designed for two dimensions becomes clumsy and time consuming. Simply navigating around a three-dimensional environment can be difficult. New devices such a 3D mouse can help, but still required a new skill and a high cognitive load to operate. The difficulty of working in a 3D virtual environment is often reported as a barrier for expanding its use in engineering design.

b) Observation in 3D

The normal method of rendering a three-dimensional world onto a two-dimensional display monitor is akin to looking at a picture or through a window into another realm. As with traditional pictures of objects or environments like landscapes, some understanding of the context is necessary since the picture is likely to be smaller than how it would look in reality. Then there is the compounding issues of the zoom and perspective which also alter the 3D image and how it needs to be interpreted. Projections onto flat 2D screens also distort the view.

In many complex ship-board environments such as an engine room, it can be difficult to understand the scale and size of objects in a new design environment because of the limitations of viewing them on a computer monitor, which is essentially a window or camera view of the virtual world. There is no possibility to use human stereoscopic vision when using computer monitors to help understand size and distances.

c) Rendering Limitations

A complete 3D virtual model of a ship with all its internal and external details included is a large amount of data, even for modern computing hardware. Since CAD software is optimised for fast editing, it often takes a large amount of time to open a scene or move the camera position.

3. ASPECTS OF VIRTUAL REALITY

The above factors are currently limiting the wider use of digital 3D virtual constructs to inform and promote collaboration on design and engineering. A graphic system that can be considered Virtual Reality provides a visual experience that tricks your mind into reactive as if you are in a real physical environment, even though you are just seeing a digital virtual construct. Key aspects of achieving this immersion include:

- Preserving real world concepts - What is up and down and simulated gravity constraining you onto a surface.
- Stereoscopic vision - You can use your eyes to gauge size and distance as you do in real life.
- Immersive vision - Your field-of-view is wide enough such that you do not feel like you are looking through a window or slot.
- Real time vision - The view is coordinated with your head movement without noticeable lag.
- Graphical stability - The visual display is smooth, without jerkiness, stutter or frame drop-out.
- Interactive - You can interact with the virtual environment

There are a number of technologies that have billed themselves as providing a virtual reality over the years but the value of them varies significantly.

3.1 THEATRES

Cinema type single large screen front projection is often used when many people need to view the virtual world. A three-dimensional effect is sometimes provided using shutter or polarised glasses. This is often advertised as VR; however, it is the least immersive because the field-of-view is limited and cannot be controlled by multiple viewers. While this form cannot be considered proper VR, it is included so the reader can understand the limitations of the technique.

With cinema projection the perspective is only correct for a single viewer position. It is also not possible to understand the size of objects because there are multiple scaling factors such as the size of the screen and distance of the screen to the observer. The technique also usually induces a high degree of motion nausea among the audience because of navigation in the virtual world is controlled by a third person and the aforementioned inaccuracies of the projection.

3.2 CAVES

Caves and domes using rear projection have been traditionally used as the visual technology for virtual reality, however they have a number of limitations including:

- Many projectors or screens and computers are required to provide an immersive environment.
- Rear projection is required which increases the space required for such a setup. Difficult to achieve accurate scaling.
- Near-field objects cannot be displayed or approached, as they are removed from the projection when they enter the room.
- Perspective is correct for one eye position only (at most). You cannot crouch down to look under an object.

3.3 HEAD MOUNTD DISPLAYS

The recent development of smart phones has helped to dramatically reduce the cost of technologies enabling to VR. These include small high-resolution video displays, plus small and accurate accelerometers and other components. The main advantage conferred by head-mounted displays is a heightened sense of immersion due to the total replacement of visual stimulus and synchronisation of the view to the head position and motion of the viewer. The viewer is able to move freely in space, and with good software what they see remains true to scale and perspective regardless of where they are standing or looking.

4. VR A DESIGN TOOL

4.1 GENERAL BENEFITS

- Most realistic way to experience an unbuilt design without using any physical constructs
- No need to be able to interpret a drawing
- No need to imagine, everyone sees the correct picture
- Reduced abstraction and therefore reduced cognitive load
- Cognition can be applied more productively, ie. solving the problem or improving the design.
- Realistic sense of scale and arrangement
- Sight lines can be checked
- Work areas can be checked
- People can collaborate on the design together even though they may be miles or continents apart
- No need to travel to discuss leads to more frequent communication on design issues
- Easier and immersive access to the design means experts can more easily add value to the design process

4.2 LIMITATIONS

- Requires a 3D model, which can be expensive if it is not already being created as part of the design process
- Realistic tactile interaction with the virtual environment is limited
- Navigation is unrealistic (teleporting)
- Poor implementation can cause motion nausea and migraine
- Some people don't feel comfortable using HMDs for reasons of vanity or having previously had unpleasant experiences with poor implementations

5. USES OF VR IN WARSHIP DESIGN

Many areas on a warship can benefit from better design communication and collaboration. The use of virtual reality for design reviews is particularly useful. Some applications of VR using ShipSpace that have demonstrated to be useful for warship design include the following.

5.1 ARRANGEMENTS

There are many spaces within a warship that demand careful arrangement in the concept design stage. Spaces such as operations rooms, galley and messing spaces all require a high level of teamwork and can be tested in ShipSpace to ensure optimal functionality. Other spaces, such as engine rooms are so densely packed with equipment and services, that access for maintenance and removal paths can be compromised. VR can be used to quickly check dimensions, potential interferences and serviceability. Arrangement of mooring equipment, especially with the increasing use of enclosed mooring spaces, can also benefit from using VR for optimisation and validation.

5.2 SIGHT LINES AND FIRING ARCS

Sight lines and firing arcs can be accurately investigated using virtual reality, allowing ergonomic and effectiveness optimisation. Traditionally, full-size mock-ups have been used to check arrangement of equipment in the bridge. Such constructions reveal nothing about the ability to see another vessel sailing in formation or viewing activities on open deck areas. Similarly firing arcs, particularly of close-in projectile weapons used for countering asymmetric attacks, can be much better checked within a virtual reality environment.

5.3 LAUNCH AND RECOVERY

Small boat and aircraft operations require a high degree of co-ordination and are inherently dangerous for personnel involved. Visibility between various control stations is paramount and can be optimised in a VR environment against ergonomic, technical and other requirements.

6. THE SHIPSPACE VR TOOL

ShipSpace is a design visualisation and collaboration tool developed by Stirling Labs for use in the maritime and naval design process. ShipSpace reads normal CAD files directly and displays the 3D model in virtual reality, complete with all of the detail at very high fidelity. Special techniques are used to maintain rendering latency of less than 11ms so that immersion is maintained, and surfaces and light are treated carefully to create a strong sensation of presence. ShipSpace provides a range of sophisticated tools to interrogate the model in various ways, such as reading metadata, "x-ray vision", controlling object layer visibility, measuring distances and many others. The tool enables up to 64 users to collaborate in the same virtual space (VR session).

6.1 PRESENCE: THE MOONPOOL TEST

250 subjects (6:1 male:female with a diverse age range) were taken on a VR tour of an interesting ship model. At the end of the tour they were looking into a large moonpool with a four-meter drop to the water surface. Subjects were then invited to step out above the moonpool. Approximately 80% of test subjects were unwilling or unable to make this step despite being verbally reminded that they were in an office environment using VR equipment. Subjects knew with certainty that they would not be falling into a pool of water and yet were unwilling to step into the moonpool.

This test demonstrates is that the level of realism provided by the ShipSpace VR tool was so convincing that the amygdala, the small part of the brain that provides survival instincts, is over-powering the front cortex, the large part of the brain responsible for reasoning and higher cognitive functions. The authors suggest that this result supports Arnheim's theory of visual thinking. Seeing and perceiving are essentially the same process and have low cognitive load. People can then apply cognition more usefully to solving problems or optimizing solutions when they are able to see the design in VR.

6.2 PRODUCTIVITY TEST

375 subjects (heavily biased toward male due to population selection, ages ranging from 18-64) that had not previously used any VR tools were given a short 3-minute demonstration on the basic ShipSpace human interface, followed by 5 minutes of verbal instruction on specific usage and then verbal instruction as necessary. The subjects were observed during their usage of the system and asked to perform simple tasks using the virtual tools.

It was found that 50% of first-time users are capable and confident in using ShipSpace within 15 minutes, while 80% take about 30 minutes and 90% within an hour. Several instructors were used, and various approaches were trialled. The effectiveness of the instructor had a negligible effect, although the script used was seen to improve the time-to-learn by up to 30%.

A strong correlation was found between the amount of tuition required and the relative youth of the subject. It is unknown if this effect is greater than a standard age-related learning rate. Some have postulated that the effect might be due to greater exposure to virtual worlds through video games, however this aspect was not investigated and is suggested as a further avenue for study.

6.3 SIZE & DISTANCE ESTIMATION TEST

30 subjects (5:1 male: female, ages 24-58) were provided with a 10-minute familiarisation tour of a VR environment. Attention was drawn to size and distance during the induction, but no specific distances or sizes were provided for correlation. Subjects were then asked to estimate their distance to a wall and the size of a room. The ability of people to estimate size and distance in ShipSpace very similar to the baseline determined by [12].

6.4 USEFULNESS TEST

Approximately 500 subjects currently working in a field directly associated with design or engineering (most working in that role or supervising/managing others). This sample was very heavily male-biased due to the population available, with less than 5% women. All participants were familiar with the concept of VR and 30% had previously experienced VR personally. Participants were taken on a short 15-30 minute familiarization tour of a virtual environment where they were taught how to use some of the features of ShipSpace. After their tour, they were asked if they saw benefit in using ShipSpace as a design tool. More than 98% of first time ShipSpace users in design and engineering industries agreed that using VR will be a useful tool to have in their workplace.

6.5 EFFECTIVENESS OF EMBODIMENT TEST

100 subjects (heavily biased toward male due to population, ages ranging from 18-64) that had not previously used any VR tools were given a short 1-minute demonstration on the basic ShipSpace human interface, followed by 5 minutes of interactive demonstration on specific usage by an instructor embodied as an avatar within the virtual world. The instructor then provided interactive instruction as necessary.

The instructor's avatar consisted of a translucent head wearing the same model of HMD as the subject had just put onto their head, plus one or two controllers (also the same as the subject was holding). Sometimes the voice of the instructor came through headphones and to the subject seemed to come from the position of the instructor's avatar other times due to convenience the instructor was physically in the same room as the subject, so the voice appeared to be disembodied.

The subjects were observed during their usage of the system and asked to perform simple tasks using the virtual tools. After the tour, the subjects were asked if they had any difficulty accepting the embodiment of the instructor and the effectiveness of the instruction was graded by the instructor.

100% of the subjects accepted the embodiment of the instructor. The rate of learning the basic controls was marginally improved by approximately 10% but time-to-competence was much more significant, reduced by 30% compared to the Productivity Test above.

6.6 ACCEPTANCE OF HMD TEST

Approximately 750 subjects working in design or engineering fields visited an office environment and were offered the opportunity to have a tour of a virtual environment. Reluctance was expressed from 5% of subjects, most frequently citing poor previous experience or techno-phobia. When gentle social pressure was exerted and guarantees that they would not experience nausea were offered, less than 1% continued to resist (and did not experience the tour).

7. CONCLUSION

Review of research into the process of visualisation shows a high degree of correlation between a person's spatial skills and their ability to visualise in their mind objects and environments. Furthermore, good spatial skills and high cognitive loads are required to interpret drawings into mental images that are useful for solving design problems or optimising solutions. The notion of virtual reality, its benefits for design and the requirements to effectively deliver a VR experience were discussed. Subjects were tested using the ShipSpace VR visualisation tool and the findings indicate that virtual reality provides a visualisation medium that enables all people, regardless of spatial skills, the ability to understand design and collaborate more completely through a realistic and more natural human interface.

8. REFERENCES

[1] H. Helmholtz, "The Facts of Perception," in Selected Writings of Hermann Helmholtz, Wesleyan University Press, 1878.

[2] R. Arnheim, Visual Thinking, University of California Press, 1969.

[3] F. Galton, "Statistics of Mental Imagery," MIND, 5, pp. 301-318, 1880.

[4] A. Z. J. Zeman, S. Della Sala, L. A. Torrens, V.-E. Gountouna, D. J. McGonigle and R. H. Logie, "Loss of imagery phenomenology with intact visio-spatial task performance: A case of 'blind imagination'," Neuropsychologia, 48, pp. 145-155, 2010.

[5] A. Zeman, M. Dewar and S. Della Sala, "Lives without imagery – Congenital aphantasia," Cortex 73, pp. 378-380, JUNE 2015.

[6] R. &. M. J. Shepard, "Mental Rotation of Three-Dimensional Objects," Science, pp. 701-703, 1971.

[7] S. Vandenberg, "Mental Rotations, a Group Test of Three-Dimensional Spatial Visualization," Perceptual and Motor Skills 47, pp. 599-604, 1978.

[8] R. H. McKim, Experiences in Visual Thinking, Cengage Learning, 1980.

[9] J. Mohler, "Improving Spatial Ability with VIrtual Reality; A review of research and applications," WebNet Journal, pp. 28-35, 2001.

[10] H. Gardner, Frames of Mind; The Theory of Multiple Intelligences, Basic Books, 1983.

[11] S. A. Sorby, Developing 3-D Spatial Visualization Skills, Michigan Technological University, 1999.

[12] M. Strauss and J. Carnahan, "The Ability of People to Estimate Distances Along a Roadway and How Their Education, Hobbies and Occupation Affect Accuracy," in FISITA World Automotive Congress Proceedings, Budapest, Hungary, 2010.

9. AUTHORS' BIOGRAPHY

Ken Goh holds the current position of General Manager, Australia at Knud E. Hansen. He is responsible for senior mechanical engineering activities. His previous experience includes the concept design of a new submarine tender vessel using the ShipSpace virtual reality design tool and managing the detailed design of HVAC on the Queen Elizabeth Class aircraft carriers.

Robert J Spencer holds the current position of Chief Executive Officer at Stirling Labs. He is responsible for development of the ShipSpace collaboration tools. His previous experience includes development of innovative human computer interfaces at Spinfast and successfully conducting the first trial of 1 million user avatars in a single shared multi-user virtual space in cooperation with IBM & BigWorld.

Certifying the Nations Flagship – A Greater Level of Assurance

S Hunt, BAE Systems Naval Ships Ltd
S Justice BAE Systems Naval Ships Ltd

SUMMARY

This paper explores the work undertaken to produce assurance certification in support of the QEC programme and in particular the evolution of HMS Queen Elizabeth from the Ship's Staff moving onboard through the final Vessel Acceptance. It also discusses some of the procurement challenge which arose in establishing the evidence for the assurance certification.

NOMENCLATURE

A nomenclature is required for papers/technical notes using a large number of symbols, abbreviations and acronyms. SI units should be used wherever possible, and ordered alphabetically.

[Symbol]	[Definition] [(Unit)]
ACA	*Aircraft Carrier Alliance*
C of D	*Certificate of Design*
RFSD	*Ready for Sea Date*
SSLOB	*Ship's Staff Live on Board*
VAD	*Vessel Acceptance Date*

1. INTRODUCTION

On the 26th June 2017 HMS Queen Elizabeth departed Rosyth on her maiden sea trials. At 65,000t she is the largest Warship to be built in the UK. Three weeks prior to this, seven hundred members of the ships company moved onboard the ship to both work and live, the largest number of ship's staff moved onboard an RN vessel that BAE Systems Naval Ships have had to manage. How can such a large and complex vessel be certified safe from a design perspective to meet each critical milestone?
In both these key milestones, the Aircraft Carrier Alliance developed a new method of assuring that the ship was both safe to work & live on and also to proceed to sea. The ACA Engineering team developed a new system to provide an objective measure of the design intent and ship's material state. Hence, it was agreed to create a series of documents across the product hierarchy called Certificates of Design (C of D). The C of D was intended to provide assurance that the design output was correct and material state conformed to the design. This is the first time that BAE Naval Ships have followed this process for vessel in build and qualification. This paper will outline the process that evolved and will discuss some of the challenges and successes along the way. It will also highlight areas for future consideration with respect to procurement as part of this assurance process.

2. QEC BACKGROUND

The Queen Elizabeth (QE) Class aircraft carriers are the biggest warships ever constructed in the UK, each providing our armed forces with a four acre military operating base, which can be deployed worldwide. The ships will be versatile enough to be used for operations by all three sectors of the armed forces, ranging from supporting war efforts to providing humanitarian assistance and disaster relief. With its complement of embarked aircraft the QE Class will be the centre piece of Britain's military capability and will be able to operate for long periods anywhere around the world at a time and place of the Government's choosing, without relying on basing aircraft ashore in their countries. The key characteristics are shown below:

Main dimensions
- Displacement 65,000 tonnes
- Length 280 metres
- Maximum beam 70 metres
- Crew size 739

Performance
- Top speed in excess of 25 knots
- Range in excess of 10,000 miles

Propulsion
- 2 x Rolls Royce MT30 gas turbines
- 4 x Wartsila diesel generator sets
- 2 x 33 tonne propellers
- 4 x advanced induction motors

Accommodation
- Accommodation for 1,600 personnel
- Accommodation and facilities for embarked forces
- Hospital area incorporating eight bed medical suite, operating theatre and dental surgery
- Recreational facilities including fitness suites and cinema

2.1 PROGRAMME

BAE Systems is a member of the Aircraft Carrier Alliance, a unique partnering agreement between BAE Systems, Babcock, Thales and the UK Ministry of Defence, working to deliver the two QE Class aircraft carriers. The programme to build the QE Class Aircraft Carriers is a hugely significant national endeavour.

The primary ship building facilities used to build the aircraft carriers were based in Glasgow, Appledore in Devon, Hebburn in Tyneside, Birkenhead in Merseyside, Portsmouth and Rosyth in Fife where the blocks for both ships have been brought together. There are also hundreds of companies based around the UK that make up the wider supply chain and provide many of the component parts used to construct both aircraft carriers.

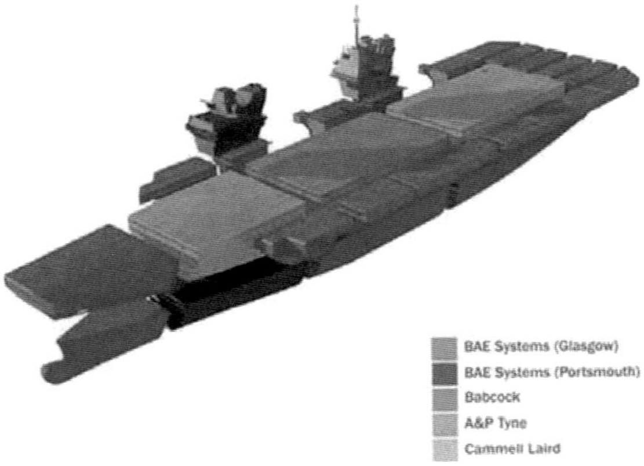

Figure 1: Breakdown of QEC Block Build

2.2 BUILD STRATEGY:

The key principals of the build strategy are below:

- Sections constructed under cover
- High degree of outfitting
- Make the most of UK industry skills and capability

This resulted in:

- Large lower block sections
- Smaller upper block and sponson sections
- Six separate build yards

Some of the key Build Strategy: Challenges were:

- Designed and constructed hundreds of miles apart
- Massive sections require transportation to assembly site
- Integrating all the sections
- Assembly and final outfit at a refit yard
- Creating one team committed and driven to deliver from multiple stakeholders, companies, sites and functions

All major block transport and assembly activities completed safely and on time. The complexity of designing and building across a number of builders and final assembly also drove the need to ensure final material standard was of sufficient maturity.

2.2 TIMELINE

Key elements of the programme timeline are shown below for HMS Queen Elizabeth (QNLZ) and for HMS Prince of Wales (PWLS). With regard to the Design, key milestones for HMS Queen Elizabeth were Ship's Staff Live Onboard (SSLOB) which was undertaken in June 2017. At this event 700 ship's staff moved onboard HMS Queen Elizabeth for the very first time. Platform Sea Trials (RFSD) commenced at the end of June, Mission System sea trials (MST) commenced in October 2017 with Vessel Acceptance (VAD) in December, 2017, being the final key milestone.

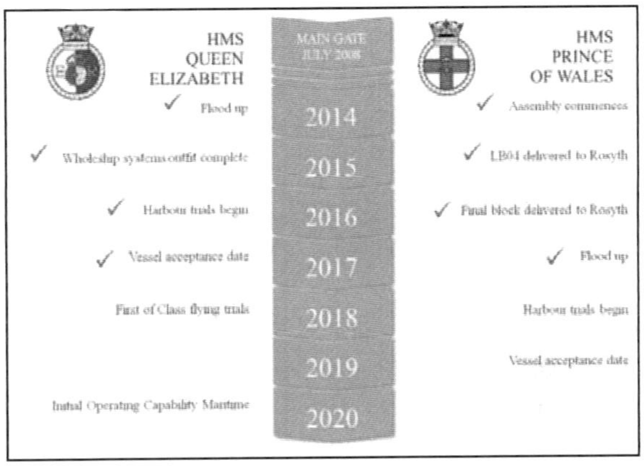

Figure 2: Programme Timeline

3. GOVERNANCE

3.1 ENGINEERING LIFECYCLE

As part of both the ACA and BAE Systems Naval Ships governance requirements there is a need to produce assurance documentation to verify system maturity and material state conformity. These are expected to be produced at key project milestones. Figure 3 below shows these typical milestones on a simplified Engineering Lifecycle V-diagram, typical for a shipbuilding project. This aligns to the programme milestones indicated in section 1.3. For QEC, an additional milestone was created for SSLOB due to the large number of crew moving onboard. For SSLOB it was necessary to demonstrate that the system required to support living onboard were at the appropriate level of maturity and availability. As the process was only initiated later in the QEC programme, the required assurance for earlier milestones was encapsulated in the first QEC milestone.

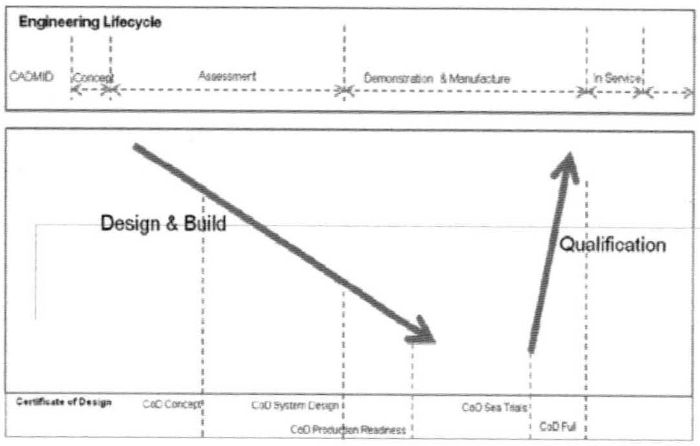

Figure 3: Typical Engineering Lifecycle

Generally, it was recognised that the previous processes of assessing a ship's material state were good but could have been a little too subjective and was not necessarily signed off by the personnel holding delegated authority. Hence design certification was required to ensure that as the QEC vessels transitioned through the various stages of the product lifecycle. The design certification confirmed that the technical solution complied with the requirements, design intent was maintained, the design was acceptably safe and that operational limitations were clearly defined and communicated. Hence it was decided to produce a suite of Design certificates, known as the Certificates of Design.

3.2 CERTIFICATE OF DESIGN STRUCTURE

Typically, a Certificate of Design is hierarchical with a top level Certificate for the whole product with underpinning Certificates of Design for key sub-systems. Given the size and complexity of a 65,000 tonne aircraft carrier this is a considerable task to undertake. For HMS QNLZ the breakdown of certification was assigned by product breakdown structure which is a clear breakdown of the ship specification and assigned ownership to the appropriate delegated design authority. This enabled a detailed breakdown to be developed with assigned ownership witch was effectively the C of D Responsibility Assignment Matrix.

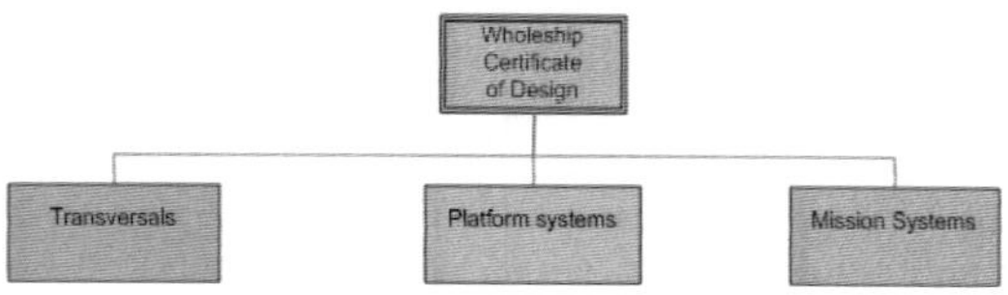

Figure 4: Certification Hierarchy (Top Level)

The top level breakdown is shown in Figure 4. This was the highest level which C of Ds were broken into, addressing the platform area (hull structure, mechanical, electrical and auxiliary systems), transverse areas and Mission Systems.

These main areas are then broken down into the system/sub system level with a certificate of design being produced for each agreed system. There is a risk of producing too many certificates so a C of D is produced at the stage where design intent can be maintained across a given breakdown. This ensures that the whole product functionality is captured by an agreed C of D. This could be achieved functionality but will be more difficult to establish ownership across discipline boundaries.

3.3 C of D CONTENT

The C of D is in itself a form of design review accounting for all the design evidence required to demonstrate that the system meets its expected requirements. The range of evidence is extensive from Design calculations, safety assessments; test forms to OEM supplied CE Marking documentation. A key part of the C of D is then the acceptance evidence, generally a test form or other verification evidence which is used to demonstrate that the material state is at a required maturity. The C of D then calls for all related safety and environmental documentation, ensuring that all hazards are referenced. A key part of the C of D is then to highlights limitations of use and identify any potential operation limitations, be they safety or performance linked.

3.4 C of D EVOLUTION

The C of Ds have evolved over four milestones now, these have been key documents in enabling the ship to pass through the various milestones. The development of C of Ds has been evolutionary as the programme moved from SSLOB to VAD, with both the number of required certificates increasing in addition to increasing maturity as shown in Figure 5 below. Furthermore at each of these events it is anticipated that the C of Ds will be at different maturity stages depending on:

- The extent of system completion required;
- The level of acceptance evidence available; and
- Level of system applicability

At acceptance, 45 C of Ds were produced for SSLOB, 81 for RFSD and 104 for VAD.

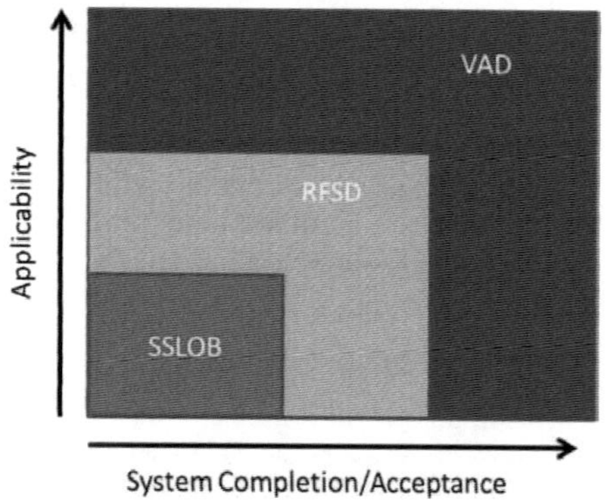

Figure 5: C of D Evolution

As a demonstration of the importance of the C of D process, certain events only occurred once the Wholeship C of D was signed off. For example the SSLOB Wholeship C of D was signed off on the Saturday morning, following signing of the last system C of D, and then the ship's company moved onboard HMS QNLZ that afternoon. Consequently the Wholeship C of D for Sea Trials was signed immediately prior to departure from Rosyth, following sign off of the remaining system C of Ds. A Certificate of Seaworthiness (COSW) is issued prior to sea trials and the Wholeship C of D is a prerequisite for signing the COSW. This proves that the design is sound, design outputs are rights and material state conforms to the design.
The Wholeship C of D was also a key part of the submission pack for Vessel Acceptance.

Figure 6: HMS QNLZ During Platform Sea Trials

3.5 OPERATIONAL BENEFITS

The C of D has grown significantly beyond its original intent of demonstrating that the ship design outputs are fit for purpose. They were latterly being used as additional safety assurance for the client and ships staff. Key system operation limitations and caveats are captured in each C of D. This information is passed to the ship's operators to ensure limitations are clearly articulated. This is vital for the safe trialling of a complex first of class ship. As shown in section 2.3 the development of system maturity progresses through each lifecycle phase to ensure a minimum level of system completion is achieved. If is therefore vital that the maturity status of system is captured. Then the operational limitations put in place. For each system where this was relevant the limitations were captured in the C of D.
The C of D process was also audited by the project's independent Safety Advisor who commended the process and also suggested that this is a "powerful process" which could expanded to include other metrics for assessing material state.
The project's client director and HMS QNLZ's MEO also commended the certification process in providing key safety assurance. The MEO advised the author that a complete pack of the information referenced in the C of D would have been extremely beneficial, unfortunately this was not practical at the current time but something for future consideration. The C of Ds were also used in further client safety briefs which the author participated in, supporting the Client Director.

4. PROCUREMENT RELATED ISSUES

The C of D relies on the submission of a significant amount of supplier or OEM data to form the evidence for sign- off. Given the stage of the project lifecycle it became evident that some of the information required from OEMs/Suppliers was deficient, either it had not been correctly supplied or had not been requested as part of the original procurement packs. This presented a number of challenges and was generally resolved after discussions with OEMS/Suppliers. However this did require additional resource/effort to resolve, particularly in the case of CE Marking, see 4.1 below. Given the gestation of the project there had been misalignment between the requirements of certification and what had been delivered by suppliers. Had the requirements for certification been established earlier of the project these issues could have been easily resolved. For future projects a recommendation will be to more closely align certification requirements with supply chain.

4.1 Conformité Européene (CE MARKING)

New Approach Directives is the term given to the EU directives that give rise to the need for conformance marking. Within the UK they are implemented through UK implementing Regulations. Compliance with these UK implementing regulations is a legal requirement. Compliance with the Conformance Marking requirements is a legal requirement and formed a key part of the C of D evidence. This was one the greatest challenges in obtaining appropriate evidence from Suppliers/OEMs generally given the lack of understanding of CE Marking and Warships. On a number of occasions appropriate evidence (being a legal requirement) was not provided and significant effort was undertaken with suppliers/OEMs to ensure that the evidence could be obtained. There was considerable confusion throughout the supply chain about a perceived exemption for Warships from CE Marking. Indeed there are very few specific military exemptions and although there are exemptions through the MED (Maritime Equipment Directive) and Classification, most electrical systems generally fall into scope with the directives, particularly with the use of COTS equipment. Again at the stage of the programme lifecycle some suppliers were more reticent in supplying this information suggesting they had satisfied their contractual obligations. Unfortunately they required reminding that this is a legal requirement and must be supplied if in scope with the directives.

For future procurement it will be sensible for engineering and supply chain to discuss these requirements as early in the project as possible to avoid having to "play catch up" and to have this clearly instructed to Suppliers/OEMs what is expected to meet legal requirements. This will avoid nugatory effort chasing down legal documentation that should have been initially supplied.

5. LR CLASSIFICATION

Whilst previous Warships have had Hull Structures LR Classification, QEC has the greatest level of LR Classification applied to a UK Warship. Complete elements of Structure, Mechanical, Electrical, Lifting Appliance and Communications have been included in the overall Class Certification. Whilst this resulted in a number of challenges to manage Class Approval was recently achieved. The Learning from Experience for HMS QNLZ is being captured and appropriate lessons learnt will be applied to HMS PWLS. These will focus on the timely management of issues and further dissemination of LR needs to those who have not been exposed to LR Classification. It is important that suppliers/OEMs are fully aware of the deliverables they need to supply to support class approval. This was particularly noticeable in the area of lifting appliances where supplier deliverables were being obtained quite late on in thc programme.

5.1 PROCUREMENT ISSUES

As with other elements of certification, the LR Class approval would also benefit from a greater understanding through the supply chain on the LR requirements. A significant number of systems require LR type approval and supporting documentation. During the recent close out of LR Approval it became apparent that some suppliers/OEMs were still not fully conversant with the LR requirements. Whilst this will be addressed for HMS PWLS, this needs further review for future procurement, particularly if Mechanical and Electrical approval will be sought.

6. CONCLUSIONS

There is no question that the development of the C of Ds greatly enhanced the ability of the ACA to provide a fully objective assessment of the platform systems design intent, verification of design output and verification that material state conforms with design. The completion of the C of D enabled HMS Queen Elizabeth to pass through each programme milestone providing assurance to the ACA, Client and the Ship's Staff. Greater alignment is required with Supply Chain for future procurement to ensure that the required evidence is available. Not only does this support the production of C of Ds but is also required to demonstrate legal compliance, with for example CE marking or its UK equivalent.

The team demonstrated a high degree of trust by the project, the Customer (DE&S) and the End User (Royal Navy). Unless and until the C of D was produced and signed by the DDA the Customer and Royal Navy were not content for Ship's Staff to move onboard or latterly for the ship to proceed to sea. This was a paradigm shift in culture as on other projects this did not happen. By producing the C of Ds it became evident from the Ship's Staff and Client Director that we, BAE Systems, could be trusted to produce a safe ship.
The team were innovative; this was a new process within a Naval Ship's project. No other project had produced C of Ds for in build activities. Through careful stakeholder engagement the various parties of the ACA bought into the process and programme for production of C of Ds. The collaborative approach in also involving the Client and Ship's Staff helped to produce important operational information which they had previously not received.
And finally the team were Bold. If they were not content to issue the C of D then it would not be issued until the relevant evidence was available. This was certainly the case where limitations were put in place which, whilst limiting a system operation, helped increase safety and reduce risk of accidents. Bold decisions were made to ensure that Safety was 1st.

7. ACKNOWLEDGEMENTS

The author would like to express gratitude to colleagues in the MoD, ACA, and BAE Systems Naval Ships who helped with the preparation of this paper.

8. ADDITIONAL INFORMATION

The opinions expressed are solely those of the author.

BAE Systems, Babcock, Thales and the UK Ministry of Defence are all registered trademarks ® of each corresponding company or government department.

9. AUTHORS BIOGRAPHY

Stuart Hunt BEng CEng FRINA, holds the current position of Engineering Manager Design Certification (QEC) and Interim Head of Naval Architecture at BAE Systems Naval Ships. He is responsible for managing Design Certification for QEC and managing the Naval Architect function.

Stuart Justice MSc CEng FIET holds the current position of QEC Technical Authority at BAE Systems Naval Ships. He is responsible for managing all technical aspects of the QEC programme.

PURPLE FIRE VULNERABILITY/LETHALITY ASSESSMENT TOOL

J S Schofield, Survivability Consulting Limited, UK

SUMMARY

Survivability management is a well-established part of naval procurement. Timely and accurate advice requires efficient assessment software backed up by expert users. This paper introduces Survivability Consulting Limited's (SCL's) Purple Fire tool, which contains unique and novel features designed to facilitate the assessments required for modern platform designs, weapon programmes and operational analysis. The tool gives analysts the ability to construct platform representations very quickly, meaning less model build time and more analysis time. It automates the consideration of large parameter spaces, allowing in-depth assessments to be conducted more rapidly than ever. A link to the University of Greenwich's maritime EXODUS is used to simulate the movement of crew through a vessel, including abandonment of the platform. The new 'criticality assessment' offers a threat-independent assessment of the quality of redundancy in a design. Case studies of the first uses of the tool are presented and the development pipeline explored.

1. INTRODUCTION

Survivability is the capability of a system to continue to carry out its designated mission(s) in a combat threat environment [1]. It is a function of susceptibility (the chance of being hit), vulnerability (the chance of loss of capability given a hit) and recoverability (the ability to regain lost capability after damage control and reconfiguration).

Vulnerability has the greatest impact of these three aspects on whole-ship structural and systems design. Therefore it is important that a vulnerability management process is in place. During the course of designing several classes of ship, an efficient and effective means of achieving low vulnerability has been established in the UK.

2. THE VULNERABILITY PROCESS

2.1 Vulnerability requirements

In order to specify that a design should be low vulnerability it is of course necessary to set numerical requirements and measure the design against those requirements. However, experience from several recent designs has resulted in a set of features that are known to produce the basis for a low vulnerability design. These are simply specified as mandated features as part of the requirement [2]. Mandated features can relate to:

- Strengthening of certain key bulkheads to protect other compartments or zones;
- Protection of high value compartments, magazines and machinery spaces;
- Shock mounting of key systems;
- Separation of propulsion machinery;
- Separation of generation machinery;
- Separation of steering machinery;
- Systems design for damage tolerance, in particular zoning of key service systems such as chilled water and power distribution, including separated redundant supply routes (pipes and cables).

Measurable vulnerability requirements relate to what level of capability is expected from the design after damage from a given threat or set of threats. More specifically, with what likelihood do we expect different functionality to be retained after a hit? The chance a given capability (e.g. ability to move at 5 knots) is available after a hit from a given threat (e.g. 1000lb bomb) is termed the Probability of Escaping Critical Damage (PECD), which of course must be set realistically – a lower PECD requirement should be set for a 20 knot capability than say a 5 knot capability, whilst the PECD requirement against a large missile will be lower than that against a smaller missile.

Requirements are typically set for the traditional "Float", "Move" and "Fight" capabilities of a platform,

The set of PECD requirements for a given type of ship design against a specific threat set is encoded in MOD policy.

2.2 Vulnerability management

Having started with a set of robust but realistic set of mandated features and numerical requirements the vulnerability of a design must be managed from stage to stage through to build and acceptance.

As each design phase introduces a greater level of detail this invariably means additional potential failure modes. The management process involves iterative checking of the design against the requirements, avoiding the introduction of vulnerabilities as the design matures and mitigation of vulnerabilities as they are identified.

The UK MOD's Vulnerability Reduction Strategy [3] is a prioritized list of measures to be applied in this management process:

- Prevent catastrophic loss (magazine explosion, sinking);
- Remove single point failures;
- Concentrate remaining critical elements;
- Separate alternative sources of capability;

- Protect remaining vulnerabilities (improve build standard, use armour or shock mounting systems);
- If all else fails don't place critical items in areas most likely to be hit.

There is nothing complicated in the above common-sense steps, but they act as a simple basis to focus the analyst and designers. Sometimes critical components or locations are obvious, but with increasing warship complexity the interplay between seemingly unrelated systems becomes important. This is why it is desirable to track vulnerability during the evolution of the design and not just at the end in an audit role. This means measuring against requirements as a design aid to identify where and why vulnerability occurs to generate recommendations and test potential design changes that mitigate vulnerability.

Quantitative vulnerability assessment is therefore a key part of the design process from the earliest stages. The biggest gains are to be made by considering vulnerability in the earliest structural layouts (hence the good sense of mandated features). The later vulnerability is considered, the more limited the scope for design change and the greater the cost of the more limited changes available.

2.3 Vulnerability assessment

2.3 (a) Overview

The vulnerability assessment combines:

- A representation of the design's structure, systems and crew;
- A representation of the threat(s) which the platform is being designed against and the means by which the threats operate (e.g. terminal trajectories);
- The science of weapon damage mechanisms, such as penetration, blast, flooding, fire, smoke underwater shock and underwater whipping;
- The numerical vulnerability requirements being measured against.

The assessment process tracks PECD for the functions on which requirements have been placed, with analysis identifying the sources of vulnerability, suggesting design improvements to mitigate the vulnerabilities and measuring their potential success.

2.3 (b) Platform representation

For effective assessments to identify all potential vulnerabilities the vulnerability model should cover:

- Hull and superstructure;
- Additional external features such as skegs, sponsons, walkways and bulwarks;
- Internal decks and bulkheads;
- Stiffening;
- Equipment components relating to Float, Move and Fight capability;
- Cables and pipes for associated distribution systems sufficient to represent all primary and alternative modes of operation;
- Crew disposition(s);
- Logical representation (as fault trees) of the range of Float, Move and Fight functions discussed above;
- Other properties such as hydrostatics, mass distribution and stability limits.

The fidelity of the model must be sufficient only for the fast-running damage algorithms built into the assessment code, thereby allowing many thousands of damage locations to be assessed in a short space of time. For instance, a particular bulkhead might be represented for blast purposes by a single panel with thickness and material specification rather than being a meshed structure as would be seen in a Finite Element model. In order for vulnerability assessments to keep pace with and guide the direction of the developing design, the platform representation needs to be constructed rapidly, within a handful of weeks, and must be modified readily as the design changes or as options for particular design features need to be compared.

Various attempts have been made over the years to build platform vulnerability models automatically from other design tools. There are a number of reasons why this has always failed or been inappropriate. The first and simplest is that at the earliest concept phases there simply isn't any other 3D representation of the design which covers all aspects of structure and the system philosophy to be included. This is particularly true when the flip-side to vulnerability assessments, lethality assessments of our weapons against target platforms, are being performed, as of course no other design representations are available.

However, even as the design develops, the vulnerability model is often the only coherent representation of structure and systems for a significant period of time. The situation reverses though in detailed design, where modern ship design tools produce 3D models of huge detail including every aspect of the platform, much more than is required for a vulnerability assessment. These can be useful for extracting aspects such as cable and pipe routes but analyst input is still required in terms of the functionality represented – a cable route is of little use without the specification of what is supplying the power and what is being supplied.

Analyst design familiarity is also gained during the platform model build process and the importance of this cannot be overstated. Many vulnerability recommendations are formed during the model build as the analyst interprets the design of each system, thus mitigations can often be made before the vulnerability assessment is even undertaken. This leaves the more intricate interdependencies of different systems as the main source of vulnerability picked up by the simulations themselves.

2.3 (c) Threat representation

The range of threats is characterised in a simple way again commensurate with the damage algorithms which will be used, in terms of their fusing, blast and fragmentation effects. Threats are often characterised in broad terms of their severity as either Design or Extreme, with different PECD targets required.

2.3 (d) Scenarios & parameter spaces

A standard assessment defines uniform distributions of trajectories on which the threat may approach the platform, on each trajectory the effect of damage (e.g. blast, fragmentation, shock, whipping, flooding) will be assessed on structure and systems. Typically for a design process port and starboard orthogonal directions are sufficient to identify all vulnerabilities, although more directions are often used for the wider parameter spaces discussed below.

Other types of scenario might consider: attacks in a specific region of the platform to compare the performance of different design options; the risk from and consequences of a weapon affecting stored munitions; or attacks likely to affect the ability of the crew to move around the platform in the case of an Escape & Evacuation assessment.

The extension to the ability to undertake vulnerability assessment of new designs is the ability to characterise the vulnerability of the fleet as a whole for wider operational analysis, whilst the flip-side of vulnerability assessment is lethality assessment to understand and optimise the capability of weapons to achieve results across a wider range of scenarios. Thus there is a variety of parameter spaces which a tool needs to address if a single consistent approach to be achieved.

Examples of recent parameter spaces assessed include:

- Design phase assessments: 10-20 threats from a small number of directions to test a design (and potential design variants) – roughly 100 grids;
- Fleet vulnerability review: 10-20 threats from a small number of directions to test each class in the fleet– roughly 3,000 grids;
- Lethality review: range of weapons against a defined target set for a range of engagement angles and firing ranges, single and multiple hits – roughly 10,000 grids;
- Terrorist threat to commercial ships: a set of generic ships of a range of sizes against a variety of sizes of improvised device – roughly 15,000 grids;
- Missile lethality: assessment of a set of weapon concepts against a range of future targets across a wide parameter space of attack angles and other scenario aspects – over 100,000 grids.

Such parameter spaces are much larger than those historically considered, which is one aspect which has driven the design and development of SCL's Purple Fire tool.

3. PURPLE FIRE

Under development since 2013, Purple Fire has been designed from the outset to facilitate the sorts of assessment required for modern platform designs, weapon programmes and operational analysis in support of the fleet. It consists of three parts:

- A data manipulation tool for the various input/output files and management of simulation parameter spaces in the form of a Microsoft Excel Add-in;
- A 3D visualisation tool embedded within the Excel window;
- A numerical simulation tool which calculates platform vulnerabilities for the required damage mechanisms and parameter space, controlled by the Excel Add-in.

An Excel add-in was selected for data manipulation because experience showed that given a bespoke tool, analysts tend to end up writing pre- and post-processors and manipulating inputs and outputs in Excel anyway! This should not give the impression that Purple Fire is "just a spreadsheet tool", far from it. Rather it should show that the tool has a very efficient method for data entry and analysis, built on a standard application used by all engineers.

3.1 Platform representation

The main aspects of the platform model discussed above are now illustrated in the form of a generic Fast Attack Craft, which is small enough to allow features to be seen at the scale of this paper.

A triangulated shell as shown in Figure 1 represents the hull and superstructure, also acting as a definition of 'inside' and 'outside' for the model. The shell is a continuous surface with no holes, these are represented separately. It can be coarsely meshed, there is no need for a very large number of triangles as seen in Finite Element models, although a 3D CAD model can be imported to Purple Fire if needed – this is straightforward, again because the data is held in an Excel sheet.

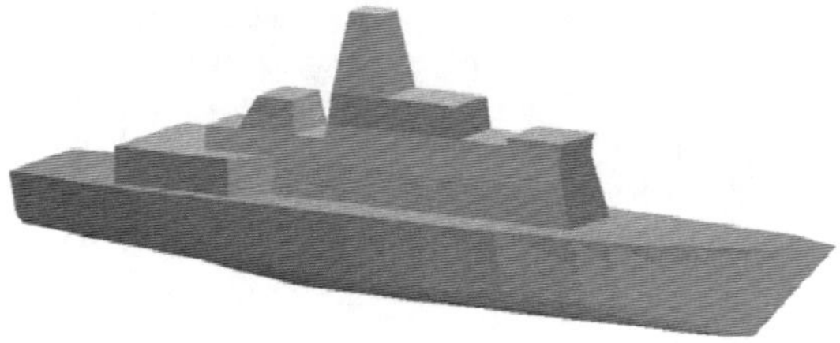

Figure 1: External structure

The main internal structure as shown in Figure 2 is made up of flat panels cropped to the shape of the external shell. The panels represent boundaries between or within compartments. Each panel has simple properties such as thickness, material, welding/joint type and insulation. Additional structural detail such as stiffeners can also be defined, which is explored as a specific benefit later.

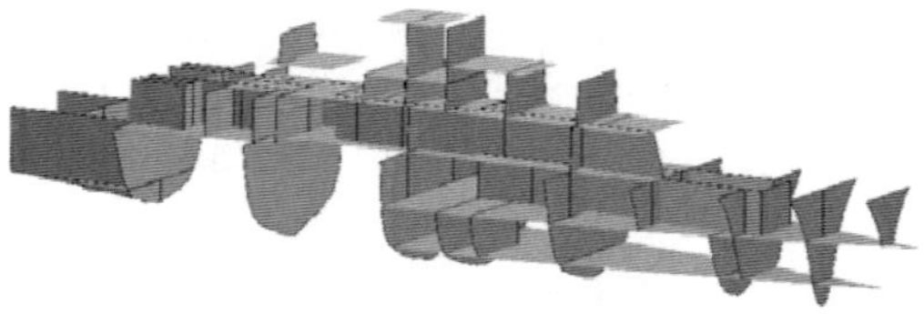

Figure 2: Internal structure

Openings within the structure as shown in Figure 3 are represented to allow for their effect on blast/fragment spread, longitudinal strength and of course for crew movements.

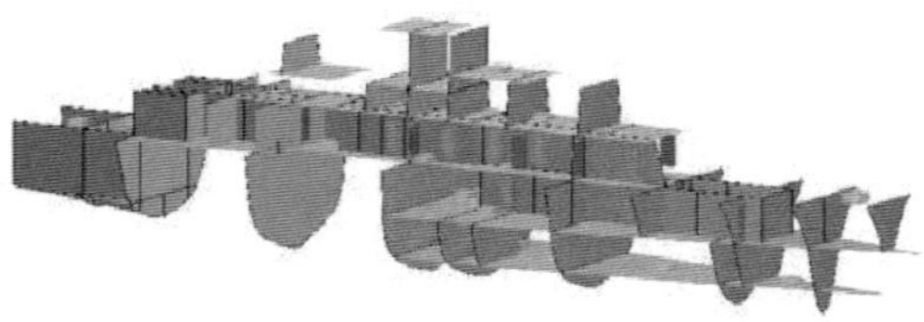

Figure 3: Openings

The functionality of the platform is also built up from simple components. In the simple example in Figure 4 various propulsion, sensor and weapon components, as well as crew members, can be seen

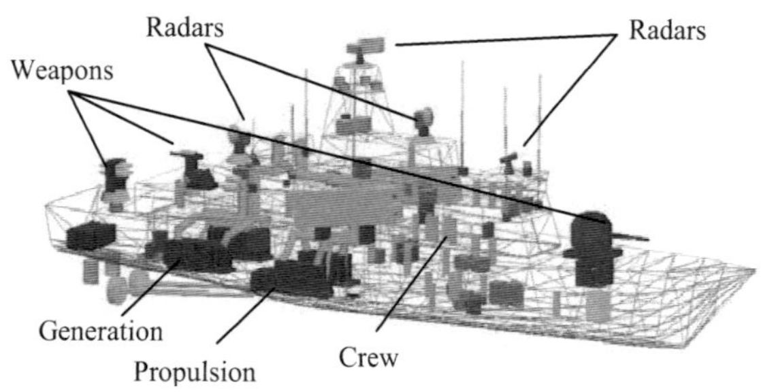

Figure 4: Equipment components

3.2 Differentiators for Purple Fire

Purple Fire's capabilities differentiate it from other available tools in a number of areas. Many of these relate to specific details of great importance to the user but not the wider stakeholder community. However, a number of substantial features are worthy of expansion.

The main strength lies in the efficiency of the above platform model building, parameter space definition, simulation and analysis, which take significantly less user and processor time than predecessor codes.

3.3 Ease of platform build

The spread sheet nature of the data allows the user to see, interrogate and manipulate multiple parts of a model with ease, duplicating and editing components in an intuitive manner. The representation of targets is generally similar to other vulnerability codes but editing of complex compartment layouts is easier and there are specific improvements such as the amount of structural detail possible. Figure 5 illustrates the explicit stiffener definition and visualisation possible, used in the simulation of underwater whipping and which also has an effect on the penetration of fragmentation through a platform.

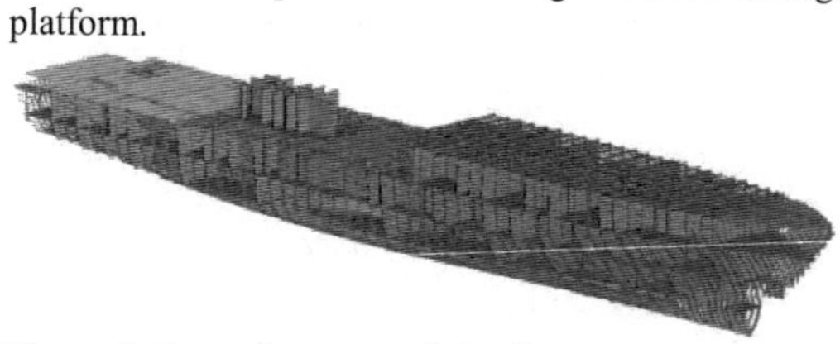

Figure 5: Example structural detail

Where the longitudinal beam of the ship is being considered, structural effectiveness and stress shadow zones are accounted for automatically in accordance with Lloyds Rules methods, producing a better fidelity of representation with much less user input required than previous codes.

The functionality built into the platform model can be efficiently displayed in the form of Block Diagrams representing the possible ways of performing a given function. For example, there may be redundancy of power supply to a given user, with power taken from one of two switchboards, each of which is supplied from a different diesel generator. This is represented by a diagram of series and parallel legs, where if a route can be traced from top to bottom without passing through an undamaged item the function is deemed to be available. The functions can very quickly become extremely complicated – imagine the diagram for a Destroyer's ability to provide AAW – but by breaking this up into small chunks and providing robust visualisation the tool helps the analyst understand the functions. An example of a real function, expanded to a limited degree but illustrating the convenient viewing of a high level of complexity, is shown in Figure 6 (where it is not expected that the text be readable at the scale of this document).

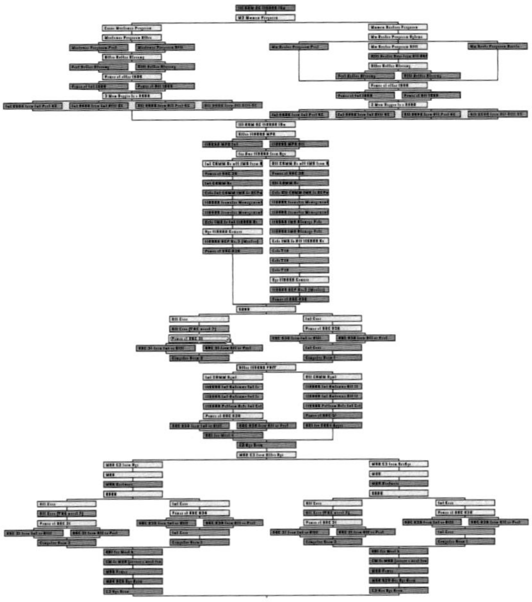

Figure 6: Example Functional Block Diagram (text not expected to be readable at this scale)

3.4 Parameter space simulation

Perhaps the biggest advance in Purple Fire over previous tools is the ease with which parameter spaces can be set up without resorting to separate pre-processors or text editing tools. A parameter space can be configured from a range of input variables, such that a single input line can represent a wide range of simulations. For example, a single entry to test a set of design variants against a range of missiles from a range of angles might be:

- Target=variant1, variant2, variant3;
- Threat=missile1, missile2, missile3;
- Azmiuth=0,45,90,135,180,225,270,315;
- Elevation=0,15,30,45;
- GridSeparation=1.

Thus a parameter space of 3x3x8x4=288 grids, each of potentially thousands of attack trajectories separated by 1m is automatically populated, simulated, and the results extracted for analysis.

Huge parameter spaces can thus be set up simply. The largest set of simulations to date was specified in just 12 entries.

The parameter space can be modified during the progress of a batch of simulations and Purple Fire gracefully handles any errors that may occur with any simulation tool, meaning that one issue on one attack trajectory on one grid cannot jeopardise the rest of the parameter space, leaving just one grid for the analyst to revisit.

Experience has shown that a standard set of a small number of types of results table are desired by the user for the range of assessments normally undertaken. The generation of these are again automated, with the user specifying a number of analysis tables, each according to what parameters should be fixed or varied, with the code instantly extracting results for analysis in consistently formatted outputs. This reduces the possibility of user error normally applicable in the manual manipulation of large data sets in spread sheets.

The combined effect of the above aspects is a saving in the effort required for assessment tasks of around 20%, higher in the case of the most complex parameter spaces, which makes vulnerability advice available even more readily in the design process and at lower cost.

3.5 Escape & Evacuation

Beyond the desire to simply build low-vulnerability warships the UK MOD has an established duty of care to its personnel including ensuring a design possesses adequate lifesaving and abandonment measures. Quantitative vulnerability analysis can also aid in this. Purple Fire links to the world-class maritimeEXODUS (mEX) code developed by the Fire Safety Engineering Group at the University of Greenwich in order to undertake E&E assessments.

mEX represents multi-compartment geometry of the vessel and the spaces within which the agents can move as a mesh of nodes linked by a set of paths. Each node represents a region of space typically occupied by a single agent (in this case a crew member).

The node network can be quite complex given the geometry of a ship, as can be seen in Figure 7. This also shows the outline of the hull and other obstacles which provides the physical interpretation of the node network. A complete ship model can be built from multiple deck geometries linked by user-definable ladders or stairwells.

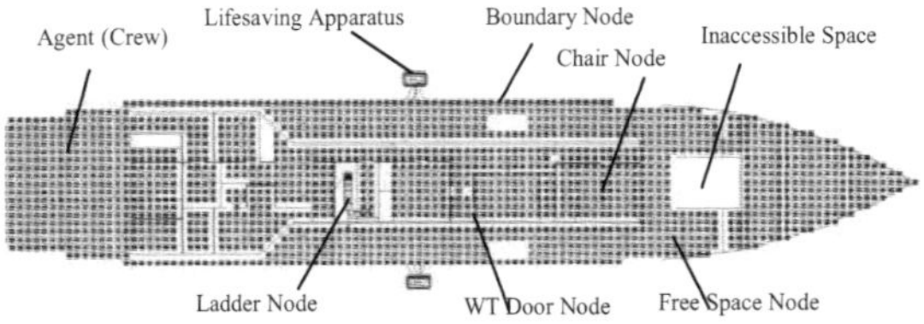

Figure 7: Example of ship geometry and associate node network

The modelling of agent behaviour in mEX is probabilistic and each individual simulation will result in different predicted timings and outcomes based on the particular agent behaviour. This includes random door opening and stair traversal times (between the maximum and minimum allowed times) and probabilistic behaviour in overtaking, congestion avoidance, etc. Consequently, simulations are run multiple times to ensure the output variables have converged to stable values.

The now-recognised benefit of this approach over legacy hydraulic tools reflects the fact that human abandonments are subject to uncertainty and cannot be realistically described deterministically.

mEX can also account for the effects of static heel and trim on agent motion, timings to open doors and transverse stairs etc.

mEX also has the ability to incorporate dynamic hazards to agents as the simulation progresses. This includes pre-determined fire hazards such as heat, smoke and toxic products which are calculated from other software tools. These act to modify the agent's ability to move through the network. It can force the agents to crawl to get under the smoke or divert to alternative routes if blocked by the hazard. Finally, agents are killed if their defined thresholds are exceeded e.g. if they are overcome by smoke or toxic products.

The link is designed to automatically launch numerous mEX simulations based on the details of the target model and the damage experienced by the threat. The link allows the user to:

- Automate the creation of an undamaged baseline mEX model from the vulnerability model;
- Automate the creation of battle damaged mEX models and extract timings back into the vulnerability assessment to create statistical E&E and recoverability metrics.

3.6 Criticality Assessment

The design process of a warship is typically time-constrained and the quicker recommendations can be made the more likely they are to be taken up. This was the impetus the development of the system criticality algorithm to provide threat independent analysis of target model's system fault trees through the identification of the critical items in the tree. Instead of using a threat with a given damage potential as during a vulnerability assessment, the criticality algorithm systematically assesses each system to determine the effect of its constituents on the overall fault tree. This method only requires a sub-set of the total data required in a target model (i.e. basic structural definition, equipment location and logical "Float", "Move" and "Fight" systems) and can be completed much faster than a vulnerability assessment.

Criticality analysis identifies the importance of the components of a system (be it "Fight" or "Move"):

- Critical Items: items which if killed disable the parent function (coloured red in diagrams);
- Redundant Items: items which if killed in pairs will disable the parent function (coloured yellow in diagrams);
- Dual Redundant Items: items which if are killed in triplets will disable the parent function (coloured blue in diagrams).

Of course, the reason why equipment criticality cannot replace lethality/vulnerability assessments is because a target's systems have both physical location as well as logical definition, as illustrated in a generic example in Figure 8. This shows the criticality of equipment for an anti-air warfare system.

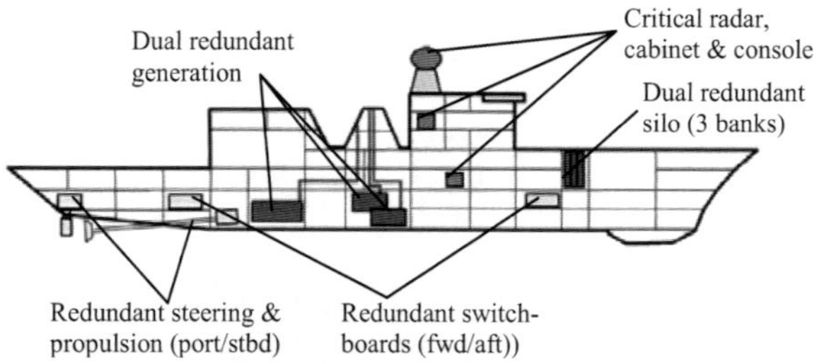

Figure 8: Equipment coloured by criticality for an example AAW system

The essential limitation of equipment criticality is evident from Figure 8 as the colouring of items doesn't identify which pairs or triplets are grouped together. Consequently, while critical items coloured red means that disabling any of these items kills the parent function, in contrast, we cannot tell which of the redundant items coloured yellow need to be killed to kill the parent function. All that can be said is that two of these redundant items must be disabled.

More useful for vulnerability/lethality estimates is the generalisation of criticality to compartments. In this case each compartment's criticality to the function is determined primarily by its own contents but also with regard to the contents of adjacent compartments. A pair of compartments whose kill will disable the function is coloured yellow, but if these compartments are neighbours then the kill may be easier, so they are upgraded to orange. This generalisation is useful since it can identify regions to target to kill the ship's function (provided that the weapon is big enough to cause damage). The generic example of this is shown in Figure 9.

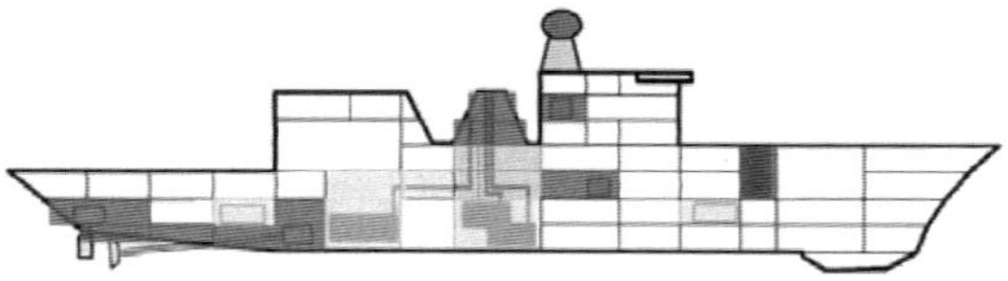

Figure 9: Example compartment criticality based on equipment contents

From Figure 9, a number of key effects are shown which are typical of criticality analysis. This includes:

- The missile silo as a whole is critical even though the individual cells are not;
- The same observation applies for the funnel;

- The aft and forward machinery spaces are redundant and both must be killed to disable the system;
- Both gearboxes (also both shafts) are in the same compartment making it critical;
- Both switchboards must be killed to disable the electrical system.

Note that, again, Figure 9 alone cannot identify which yellow compartments must be killed together without further knowledge on the system. However, just as compartment criticality can reveal which pairs of equipment must be killed, compartment pairs can be identified by extending criticality to the zone level.

The benefit of criticality assessment is that it can rapidly identify:

- A general level of robustness of the target before even assessing vulnerability;
- Optimisation of weapon targeting in lethality assessments.

4. USAGE TO DATE

Purple Fire has been used on a number of assessments, some of which are briefly discussed to illustrate the tool's growing track record:

- Lightweight Torpedo Research Phase 1, July 2015 – Requirement to establish how the performance required from a small directed energy torpedo warhead depends on hit location for a number of submarine targets. The effective shot-line length from impact point to pressure hull for a number of potential detonation geometries was calculated. The structural models used were documented, verification of methodologies was given and examples of results were presented.
- Detonation Point Boundary Phase 2, July 2015 – Conducted a Purple Fire assessment using Hicks' elastic beam methodology of whipping on three targets with and without plastic hinges caused by early time shock damage. The results were compared to establish the effect of the hinge on target motion and damage potential.
- Comparison of submarine models, July 2016 – Various assessments of torpedo lethality had been undertaken in the past. SCL were tasked to explore some of the differences between the assessments. Recommendations were made as to an appropriate damage model for use in future assessments, consisting of changes to the Parnell shock holing equation. A verification of SCL's implementation of the appropriate Parnell equations and assessments was undertaken.
- Lightweight Torpedo Research Phase 2, June 2016 – Phase 1 was extended for further assessments and additional targets to understand the largest tolerable hit cone angle that can achieve a given threshold penetration to identify the constraints on targeting capabilities to ensure a greater reliability of penetration into each target.
- Type 26, July 2016 & on-going – Purple Fire was used to generate the section properties for use in the T26 whipping assessment including automatic determination of stress shadows around discontinuities and openings.
- Future Offensive Surface Weapon Lethality, Nov 2016 – A top-level assessment of indicative performance and considerations of defeat mechanisms was undertaken for a wide range of targets through the use of system criticality in Purple Fire.
- Minimum Damage Effects Requirements, Aug 2017 – Construction of a range of red submarine target platforms.
- SVHO Escape & Evacuation, Apr 2017 – First use of SCL's link between Purple Fire and Maritime Exodus, marking the first use of the latter E&E code on a serving RN platform.
- OPV Vulnerability, July 2017 – Generation of sectional data from full structural definition in Purple Fire for input to whipping assessment. E&E assessment using link between Purple Fire and Maritime Exodus.
- Future Offensive Surface Weapon Lethality Targets, May 2018 – Construction of a range of red surface target platforms.
- Type 31, on-going – Construction of Purple Fire vulnerability models of a number of candidate designs to facilitate assessments in existing code, more advanced whipping and E&E assessments and to future-proof modelling process.
- Variation of vulnerable regions, on-going – Criticality assessments of a number of existing Purple Fire targets to advice on potential targeting philosophies. Assessment of a parameter space of weapon concepts against the targets.

5. DEVELOPMENT PIPELINE

Purple Fire already has unique features and efficiency, however a number of developments are planned to further improve functionality.

5.1 Probabilistic magazine assessment

Purple Fire currently undertakes a traditional deterministic magazine reaction simulation, whereby the stimuli from damage mechanisms are interpreted as a deterministic status of each stored munition (e.g. burn, deflagrate, explode, partial detonation, full detonation) which then leads to subsequent additional damage. This does not reflect the full detail of SCL's probabilistic methodology used (through fairly manual means) in Type 26 and Fleet Tanker assessments and accepted by the Naval Authority Group. In this method, the likelihood of damage to a munition (for example there may be a 60% change of fragment impact to a warhead) is taken into account and all potential reactions are simulated. Purple Fire has been designed to accept this methodology, which might for instance mean that one initial attack automatically becomes two different scenarios, the first with no reaction and weighting 40%, the second with a reaction and weighting 60%. Subsequent waves of reaction make the possibilities even more complicated, but this can be catered for by an iterative addition of subsequent events to each scenario.

5.2 Residual strength & effectiveness of superstructure

Purple Fire already calculates all of the sectional structural information needed for an assessment of residual strength (since very similar information is needed to build the beam model for its whipping assessment). It is intended to add a residual strength calculation, along with the ability to examine the effectiveness of different areas of structure including superstructures, which can be subject to uncertainty.

5.3 Reactive Materials

SCL is under contract with an industry partner to develop and implement in Purple Fire an algorithm for the penetration and other aspects of reactive materials used in warhead casings.

5.4 Improved UNDEX methodology

SCL has worked with an industry partner for Dstl on the initial stages of an improved methodology for the combined effects of shock, bubble jetting and whipping. Purple Fire is designed to accept the expected methodology which will be implemented if the work progresses.

5.5 Recoverability

SCL has undertaken initial design of a methodology to produce useful probabilistic measures of recoverability during the concept and design phases with a quicker and less detailed approach than more deterministic treatments.

5.6 Detailed Fire

The current simple fire method is not sufficient to simulate the effects on crew, including crew movement, of fire and smoke. SCL intends to implement a more detailed simulation option, likely by linking to an external tool such as SMARTFIRE from the Fire Safety and Engineering Group (FSEG) at the University of Greenwich which runs alongside their Exodus capability already included as Purple Fire's evacuation and crew movement algorithm.

5.7 Detailed Flooding & Stability

Although not a current priority, it is intended to add an iterative flooding and stability check to Purple Fire to simulate the time-dependent flooding and attitude changes following damage, linking these to E&E assessments.

6. CONCLUSIONS

Policy and simulations have ensured that low vulnerability has been designed-in for all recent classes of ship for the Royal Navy and Royal Fleet Auxiliary. An efficient assessment tool is essential for the necessary analysis to be achieved in the required timescales. Parameter spaces and the complexity of analyses have been increasing over time. Purple Fire has been designed to facilitate such assessments, reducing previously laborious processes, improving the quality of results and reducing the chance of errors. Thus it allows faster, more advanced assessments for reduced cost.

7. REFERENCES

1. NATO, ANEP43 Ship Combat Survivability 2nd Edition, September 2002.
2. DE&S, Key Threshold Requirements for Surface Ship Vulnerability, Version 1, December 2015.
3. DE&S, MAP 01-105 The Vulnerability Reduction Compendium, Issue 2, March 2015.

8. AUTHOR'S BIOGRAPHY

James Schofield is Managing Director of Survivability Consulting Limited. He is the lead developer of Purple Fire. His 17 years of experience in naval vulnerability and lethality includes vulnerability assessments of Type 45, Queen Elizabeth Class, Type 26, Fleet Tanker and Fleet Solid Support ship.

EXPERIMENTAL INVESTIGATION OF FLOW OVER THE FLIGHT DECK OF A GENERIC AIRCRAFT CARRIER

K Vignesh Kumar[1], **MP Mathew**, **SN Singh**[1], **SS Sinha**[1] and **R Vijayakumar**[2]
[1]Department of Applied Mechanics, Indian Institute of Technology Delhi, New Delhi, India
[2]Department of Ocean Engineering, Indian Institute of Technology Madras, Chennai, India.

SUMMARY

Flow structures and air wake studies are crucial features which need thorough understanding in the initial stages of design, to enhance the primary role of aircraft carriers- to launch and recover aircraft, safely and swiftly. Hardly any information is available in the open literature which could be used to establish the important flow features of the carrier environment. Any attempt to establish such a benchmark requires to be generic, repeatable and scalable. The benchmark data would be useful for validation of numerical models. The relative ease of setting up CFD simulations compared to physical experiments can consequently be exploited to undertake numerous parametric analyses to evaluate design alternatives.

This paper presents experimental studies carried out in wind-tunnel at IIT Delhi to measure the airflow over a generic aircraft carrier model and establish baseline data of the pressure distribution over the flight deck. Several pressure-taps embedded over the deck give the pressure at various locations which are extracted for data analysis. Numerical simulations have been validated against the experimental data and a suitable turbulence model has been identified for the problem. These studies offer valuable insight into the flow conditions over the flight deck and provide a logical start point to extend the work to qualify multiple geometries and locations of the island for optimizing airflow. The explication of flow physics thus obtained is a prerequisite to understanding the influence of these structures in the critical 'burble' zone of aircraft landing approach.

NOMENCLATURE

V	Free stream velocity (m s^{-1})
p_∞	Free stream reference pressure (N m^{-2})
p_i	Local pressure at i^{th} location (N m^{-2})
ρ	Density of fluid (kg m^{-3})
C_p	Pressure Coefficient
x	Distance from ship's bow (m)
L	Length of Ship model (m)

1. INTRODUCTION

Naval aviation is a fascinating and complex field of military operations. It combines two powerful assets- the aircraft and the warship, to create a formidable force multiplier. The versatility, increased range of action and potency that this combination derives, comes at the cost of increased complexity and augmented risks. Compared to land-based aviation facilities, aircraft operating from ships at sea are burdened with constraints of space and unrelenting environmental conditions.

The epitome of ship borne flight operations is the launch and recovery of aircraft from an aircraft carrier. The 'carrier' is a seagoing airbase fully equipped to supply, arm and fly fixed-wing planes and rotary wing helicopters from its flight deck. But in striking contrast to an airbase, the carrier has to be equipped with the entire array of flight equipment, in less than a tenth of the space, while it sails at speeds nearing 30 knots, in sea states which are beyond its control.

Landing an aircraft on a carrier ranks as one of the most demanding tasks for naval pilots. They are required to approach for landing at speeds over 200 km/h (about 55m/s) with an available runway length of only around 100m, with other possible obstructions (equipment, personnel, etc.) positioned within just a few meters of the wing tips. The ship's continuous sea keeping motions deprive the pilot of a stationary platform more suited for landing. The performance limitations imposed on the aircraft due to its wings' high angle of attack during landing, adds another layer of complexity. Moreover, the limited reaction time available to negotiate these challenges increases the pilot workload tremendously.

1.1 THE BURBLE EFFECT

Over and above the complexities described in the previous paragraph, one significant factor that raises the workload of the pilot is the requirement to land, crossing the wake of the carrier. The luxury of maintaining considerable separation distance from features in the vicinity (that is possible in a well laid out airfield) is not available to the pilot approaching the carrier. The superstructure (island), deck/hull edges and other physical features generate turbulent air wake characterized by regions of large separated flows. In aviation parlance, this zone of disturbed flow is called the 'burble' and is encountered by pilots during their landing approach [1].

The 'burble effect' has been reported by pilots to occur mainly within the final half mile (about 800m) to touch down. The interaction of the wind on deck and the ship's 'bluff' structures causes the flow in the wake of the carrier to mainly manifest reduced wind velocities and a downward flow direction (due to the hull geometry), which in turn reduces the lift generated by the wings resulting in a sinking effect. The pilot is required to compensate using thrust and controls to maintain the glide slope. The limited time and distance available for the pilot to commit to the landing at this stage of carrier proximity makes the burble a significant cause for carrier landing error.

1.2 EFFECT OF BURBLE ON PILOT WORKLOAD

The task of landing on an aircraft carrier requires precise airplane control by the pilot since the tail hook must engage one of the four arresting wires spaced only about 12 meters apart. An aircraft approaching at 200 km/h (55 m/s) along a 3.5-degree glide slope will nominally clear the ramp by 3- 4 m and touch down within the next second. Any change in the vertical height during the approach of the aircraft will cause the pilot to overshoot/undershoot the targeted touch down point on the deck. An overshoot dictates that the pilot aborts the landing attempt, 'bolter' and take off performing a 'touch and go' procedure to become airborne again. An undershoot, on the other hand, may be catastrophic, as the aircraft may strike the stern of the carrier resulting in an accident/crash causing damage not only to the inbound aircraft but also to the ship, rendering the landing strip unusable.

The problem is amplified further because the pilot has only between 1-2 seconds to commit to/abort the landing. Such split-second decision making imposes an enormous strain on the pilot given the severe consequences that pivot on his decision. The situation could get further aggravated because of ship motions, bad weather, poor visibility, strong winds and the level of turbulence.

Therefore, it emerges that pilot workload is attributable principally to three causal factors namely ship dynamics, aircraft dynamics, and pilot response. The latter two will inevitably be affected by the disturbances in the flight path due to the 'burble,' and hence any endeavour towards its mitigation will positively reduce the pilot workload, thus enhancing safety and reducing landing accidents.

2. MOTIVATION

The prime motivation for this study is the safety of life. Pilots embark on the hazardous task of operating from the carrier at the peril of their lives, and therefore research towards making the process even marginally safer is propitious. This section analyses the causes of aircraft carrier landing accidents and the motivation behind investigating the burble as a means to enhance safety by incorporating it as a design driver.

2.1 CARRIER LANDING ACCIDENTS

A study by Hoiberg [2] indicates that fixed wing carrier pilots experience long-term medical ailments as an effect of flying carrier missions. They exhibit higher mortality and hospitalization rates for several disorders compared to other pilots. Most of the deaths recorded in the study were attributed to forced landings. Analysis of aircraft carrier landing accidents by Bricston et al. [3] cite pilot error as the most common cause, accounting for about 78% of accidents in their study. Goorney [4] conducted studies for the Royal Air Force to analyse the human factors involved in the pilot error and concluded pilot fatigue as one of the chief reasons for carrier accidents. Collicot and Muir [5] correlate temporary mental overload imposed on the pilots during such high-risk carrier operations as a critical factor for pilot error. Although the carrier induced turbulent burble is not solely responsible for landing accidents, it undoubtedly adds to pilot fatigue in the perilous final few seconds of landing. For instance, the characteristic 'downwash' velocity created by the burble demands rapid control inputs by the pilot to maintain his vertical flight path along the glideslope. Durand and Teper [6] report that approximately 80% of all landing accidents in the US Navy and 25% of bolters correspond to the lack of flight control in the vertical plane. These facts encourage investigation of the flow characteristics over and behind the carrier and to determine how the conclusions derived from this study can be incorporated to drive future designs.

2.2 AIR WAKE ANALYSIS AS A DESIGN DRIVER

The flight deck is at the very core of aircraft carrier design. Primary design drivers such as the number and size of aircraft dictate the size and shape of the flight deck. The merit of a flight deck configuration is typically quantified through the use of metrics such as sortie generation rate and aircraft turnaround time. The position of parking of planes, pre-launch, post-recovery and for maintenance/ servicing purposes considerably affects these metrics and thus the layout of the flight deck. Ship designers, typically do not consider the aerodynamic aspects of the various elements on the flight deck and its effect on aircraft operations on-board the carrier during the early design stage.

An understanding of carrier air flow structures is imperative and is an essential design aspect. Ability to factor the air wake characteristics as a design driver and to modify the design in its early stages to obtain a more favourable wake field in the flight approach zone would be immensely beneficial. Such an ability to predict the air wake over a carrier can then be used for early 'diagnosis' of structures that would create an unfavourable wake and alter them accordingly. This knowledge ultimately will lead to enhanced safety, reduced pilot workload and lead to lesser bolters and hard landings.

2.3 REQUIREMENT OF IN-HOUSE RESEARCH

A detailed review of the published literature on the subject is presented in the next section. Though multiple studies have been carried out on 'burble' turbulence behind an aircraft carrier, most of these studies pertain to specific individual aircraft carriers in active service/ under development. Hardly any information is available in the open literature which could be used to establish the important flow features of the carrier environment. Any attempt at establishing such a benchmark requires to be generic, repeatable and scalable. The benchmark data would be useful for validation of numerical models. The relative ease of setting up CFD simulations compared to physical experiments can consequently be exploited to undertake numerous parametric analyses to evaluate design alternatives.

Hence, there is a requirement for in-house research on the fundamental aerodynamic aspects of an aircraft carrier with particular relevance to the aspects of flight deck design (including the island structure), which contribute to the turbulent wake behind a carrier

3. REVIEW OF LITERATURE

The available published literature pertaining to the subject is described under three categories. The first section presents literature based on predominantly experimental work. The second section outlines literature involving numerical CFD simulations and also empirical/ mathematical models. The third section brings out various flow improvement devices and methods proposed to improve carrier air wake.

3.1 EXPERIMENTAL STUDIES

Ever since 1920's when the first-generation aircraft carriers sailed the seas, the effect of airflow around aircraft carriers on flying operations has been a topic of interest and concern. During the 1960's numerous studies of air flow patterns over and about various carriers in service, had been carried out in both full scale and model scale. It was during this period that the perspective of such air flow studies changed from surveys on existing carriers towards considering the air flow in the design stage itself and incorporating changes in the configuration to improve the air wake of future ships. Among the first set of experiments to study this problem were those performed by Olenski and Dankworth [7], Kjellman and Colt [8], Hanegan and Badger [9], Oldmixon [10], and those presented in the US Navy Reports [11] and [12]. These experiments offer some insight into the flow characteristics and evaluate dynamic parameters associated with aircraft landing on carriers. These were the first set of studies that considered different wind-on-deck conditions. Barnett and White [13] conducted experiments to compare the airflow characteristics over several aircraft carriers. This work was an attempt to derive correlations between carrier geometry and air wake and extend the idea to improve flow patterns over future designs.

Durand and Teper [6], formulated a mathematical model considering deck motions and turbulence inputs for terminal flight path control. Lehman [1], conducted experimental studies of the burble by studying a Forrestal Class carrier in a water tunnel using cavitation techniques using a mercury water manometer for measurements. He reiterates that the wake field obtained in static condition is vastly different from that obtained with the ship undergoing dynamic motions, however the static study does offer an opportunity to study the changes in the gross wake field with ship orientation. He identified that the leading edges of the bow and landing ramps can shed vortices similar to 'wing tip shedding' from an aerofoil. He also noticed that since the island structure is at an angle of attack to the flow, it creates disturbances in the aircraft approach path acting as a bluff body placed at an angle of attack. He assumed that an extremely detailed model would not be required as only major features of the ship geometry would produce disturbances large enough to affect the heavy aircraft borne on such carriers.

Rajagopalan et al. [14], conducted experiments on a 1:48 scale Landing Helicopter ship in a wind tunnel using PIV for measurements. The results were compared with and without factoring the wind tunnel walls with numerical simulations. It was found that the inclusion of the walls had minimal effects on the flow field and up wash velocities in the regions of interest.

The more recent experimental study conducted by Cherry and Constantino [15], aimed to qualitatively and quantitatively model the air wake created by an aircraft carrier flight deck and superstructure in order to understand how it affects aircraft on approach and landing. The experimental approach involved using a five-hole Pitot probe rake system for mapping of the air wake of a 6 feet aircraft carrier model. Wind tunnel tests were performed at Reynolds numbers of 1.1×10^7 and wake maps of 288-point measurements each were generated for configurations with and without the superstructure. The experiments have shown that the superstructure created a region of extremely low flow velocity and downward flow angle. While the hull/deck configuration produced velocity deficits as low as 17%, the presence of the superstructure increased this to almost 30%. Furthermore, the addition of the superstructure decreased the downwash angle at the deck edge from -5 to -8 degrees.

3.2 EMPIRICAL/MATHEMATICAL MODELS AND NUMERICAL SIMULATION STUDIES

For the purposes of studying/simulating the burble, an engineering model of the disturbances is required. This engineering model may be the simplest or minimum acceptable model that correctly identifies the primary parameters of particular interest. The most widely accepted and used model for simulation of the carrier burble is depicted in the Military Specifications, MIL-F-8785-C [16]. Steinberg [17], proposed a fuzzy logic-based aircraft carrier landing system and incorporated the above disturbance model to factor the carrier wake in his algorithm.

The Department of Defense Handbook [18], based on pilot experience, reaffirms the MILSPEC disturbance model as acceptable to be used during simulation of flight approach. The MILSPEC disturbance model describes the steady components of the carrier air wake which consist of a reduction in the steady wind and a predominant up wash aft of the ship which are functions of range aft of the ship center of pitch.

The existing Carrier Approach Criteria (CAC) of the US Navy, places constraints on various parameters during the aircraft landing such as the aircraft approach velocity etc. It was believed that beyond a certain approach speed of 72 m/s (140 knots), the accident rate increases drastically. But by analyzing Naval Safety Center data from January 1980 through May 2001, Rudowsky [19], concluded that there is no longer a credible correlation between mishap rate and approach speed within the scope of aircraft reviewed and therefore should not be used as an indicator of safety. So, any increase in the apparent approach speed, if any, as a result of our endeavor to provide a more favorable air wake, is considered safe and acceptable even if it exceeds the upper limit prescribed by the CAC.

Polsky and Bruner [20], studied the air wake on an LHA ship USS Saipan after validating the results with wind tunnel tests and full-scale tests. A seven-hole probe was used to collect three-component velocity data. It should be noted that a seven-hole probe records time-averaged values of velocity. Therefore, unsteady aspects of the flow were not detected. More recently, the burble has been modelled in CFD by Polsky and Naylor [21]. They found that the geometry of the stern had a significant impact on the air wake qualities along the aircraft approach path.

Shipman et al. [22], investigated the sensitivity of the air wake solution to several modelling parameters, including geometric complexity and the resolution of boundary layers, with the aim of determining the level of fidelity required to obtain an accurate solution. Results are compared to wind tunnel experimental measurements. The results of these studies show that, in general, a majority of the air wake flow features are characterized by bluff body vortex shedding from the larger geometric entities that comprise the ship geometry. Depending on the requirements and intended use of the solution, a certain trade-off can be reached between solution turn-around/grid generation time and solution accuracy.

The most recent work in CFD studies pertaining to carrier air wake is by Kelly et al. [23]. Numerical simulations of the ship air wake have been conducted for inclusion in the flight simulators for HMS Queen Elizabeth. This ship has a twin island configuration and the flow on the flight deck and between the island structures is studied. The results from Kelly et al. [24], attempts to integrate the findings of the study to inform the flight training simulator for the carrier.

3.3 FLOW IMPROVEMENT DEVICES AND METHODS TO MODIFY AIR WAKE

Various studies have been conducted to propose modifications to the aircraft carrier geometry to modify the air wake. Several active and passive flow improvement devices have been considered for inclusion in new carrier designs or as retrofits to existing carriers with an aim to improve the air flow in the flight approach zone. This section describes literature related to such endeavors to create a more favorable approach path for the approaching aircraft.

Meier and Gale [25], have explored various alternative layouts of the flight deck geometry to ascertain if any other variations of the existing carrier geometry, would lend better flow characteristics. Their study was not limited only to the wake, but included other practical and operational aspects like parking area, flight landing and take-off areas, ease of command and control etc. The study concluded, that for the heavy fighter aircraft intended to be pressed on service onboard carriers, and for a minimum economically viable size of the air-wing, the existing layout of carriers (catapult assisted take off on a straight run way and arrested recovery on an angled landing ramp) is the best solution.

Ringleb [26], suggested various modifications to the island to get a more favorable air wake in the flight approach zone. Allen [27], explored the option of including a passive device on the carrier stern. The air passing along the flight deck is provided with an upward component of velocity resulting in an up wash in the air wake behind the aircraft carrier. The up wash persists for a reasonable length downstream of the aircraft carrier and hence, for a given spatial attitude; the angle of attack of the aircraft relative to the airstream is increased. This means that the aircraft may be flown at a reasonably high angle of attack to the airstream on a landing approach without sacrificing the pilot's view of the deck.

Nangia and Lumsden [28], list various flow improvement devices that may be installed on aircraft carriers to improve the air wake. They investigated the effectiveness of Columnar Vortex Generators (CVG). Results are presented from theoretical studies on isolated CVG at different sweep angles and then attached to a deck-side-edge. It is shown that an isolated CVG has a pronounced trailing wake at low sweep conditions. This detracts from the ability to capture inner vortical flow. Swept CVG improves the flow capture abilities. When attached to deck-side-edge, the CVG leads to an improvement of flow-field over the deck

Columnar-vortex generators (CVG) have also been proposed by Lamar [29] and Landman et al. [30], as means to increase the safety of take-offs and landings of aircraft on aircraft or helicopter carriers and other ships at sea. According to the proposal, CVGs would be installed at critical edge locations on ships to modify the vortices in the air wakes of the ships. The desired effects of modifications are to smoothen airflow over take-off and landing deck areas and divert vortices from take-off and landing flight paths.

Bow Flap was added to the forward end of an LHA ship by Czerwiec and Polsky [31] to study the improvement in flow. They concluded that the provision of the bow flap mimics a 'rounded' deck edge and improves the flow over the flight deck as compared to a sharp right-angled deck edge.

Munguia et al. [33], evaluated alternating current dielectric barrier discharge (AC DBD) plasma actuators. These were used to leverage their known ability to add momentum to the flow and thus reduce vortex shedding from bluff bodies. This project proposed that by reducing vortex shedding, the AC DBDs would attenuate the downwash in the burble. They concluded that these devices mitigate the downwash to some degree.

4. CHALLENGES INVOLVED IN THE STUDY

The literature review clearly brings out that the aircraft carrier air wake has been constantly an evolving field of study, ever since aircraft took off from the deck of the very first carrier. The 'burble effect' and its adverse effects on the approaching aircraft have been acknowledged as a significant contributor to pilot workload and resulting accidents on board aircraft carriers. Several researchers have studied the problem and made significant contributions to our current understanding of the same.

However, scrutiny of the existing literature exposes certain challenges in comprehensively studying the problem of carrier induced turbulence. These are brought out in the following paragraphs.

4.1 LACK OF FUNDAMENTAL TREATMENT

Most of the reported literature approaches the problem from a focused perspective. Studies have been limited to specific individual carriers and the effects of the wake thereof. It emerges that the problem has not been accorded a more generic treatment to identify and understand the fundamental nature of the problem.

4.2 NON-EXISTENCE OF BENCHMARK MODEL

One of the main challenges is the non-existence of a standardized, generic 'benchmark model' for carrying out experimental and numerical studies on aircraft carriers. Such a model would be extremely beneficial to study, validate, compare and collaborate research related to flow past aircraft carriers.

4.3 CLASSIFIED NATURE OF MILITARISTIC DATA

Among the literature available, numerous studies that have been carried out remain classified owing to security concerns and sensitive nature of such military data. This is primarily because these studies were carried out on (full scale/model scale) of actual naval ships in active service/intended for future service. As a result, there is very little data in the open domain, regarding air wake of carriers, to base future research upon.

4.4 IDENTIFICATION OF CONTRIBUTING FACTORS

The next challenge is to identify which, and to what extent, the major features of the carrier contribute to the problem. Although, there have been attempts in the literature to identify these contributors, there is no literature that quantifies the contribution of each of these geometrical features such as island, deck geometry, shape etc., towards the 'burble effect'.

4.5 SENSITIVITY OF BURBLE TO CHANGES IN CARRIER GEOMETRY

There is also a requirement for a systematic parametric investigation on the sensitivity of the burble to changes in carrier geometry. This would provide valuable design guidelines for future carrier development programs to follow, to obtain a more favourable air wake. At present, there's no such data which quantifies the changes in the air flow over the flight approach zone vis-à-vis the changes in the carrier geometry. This would provide a useful early design tool to optimise aerodynamics of the carrier.

5. RESEARCH PROCEDURE

Having identified the challenges posed, we declare the objectives to overcome these challenges and devise a framework for the research. This section clarifies the intention, scope and extent of the study and details the methodology to be adopted.

5.1 PRIMARY OBJECTIVES

The overlying objective of this research work is to improve the air wake in the flight approach zone, by design, and ultimately reduce the workload of the pilot approaching to land on a Carrier. The present study, in order to generalise the problem and adopt a fundamental approach, will use a simplified aircraft carrier model. To further simplify the problem and provide a logical start point for further studies, the present approach, removes the aircraft from the consideration entirely. Thus, eliminating variables like type, size and speed of the incoming aircraft, while acknowledging the fact that different aircraft and different pilots will react to the burble differently.

The geometry of the carrier, heading into the wind at some relative speed, is responsible for the burble in its wake. It is a 'map' of this air wake that this study intends to capture, assuming no aircraft is approaching. This 'wake map' is to be surveyed in the intended approach path of the aircraft and then, attempts to be made to modify and 'improve' it by suggesting changes to the geometry of the carrier.

It is anticipated that the burble may not be completely eliminated, even with the modifications intended in this research. The task is to study if the carrier geometry can indeed be modified to provide a relatively more favourable approach path for the aircraft. This would aid in setting up certain design guidelines to optimise aerodynamic aspects of the flight deck in preliminary stages itself. How different aircraft will react as they enter the burble region, is a scope for future, more detailed studies which will require to include the coupling effects of the aircraft with the carrier wake.

5.2 RESEARCH METHODOLOGY

The present study sets the course for a systematic research procedure to tackle the problem. The sequence of tasks to be carried out is listed below.

- Establishment of a benchmark geometric carrier model.
- Generation of baseline experimental data and validation of numerical models.
- Measurement of flow parameters in the wake of the carrier along the flight approach zone.
- Quantification of contribution of island, hull and flight deck geometry to the burble.
- Parametric Investigation of effect of changes in Island size and location, changes in deck geometry etc.
- Derivation of useful design guidelines from the study.

This paper focuses on the first two tasks, i.e. conceptualization and development of the benchmark model and generating baseline data of the pressure distribution over the deck of the model. This data is used to validate numerical models as explained in following sections.

6. CONCEPTUALIZATION OF GENERIC AIRCRAFT CARRIER

Research on helicopter-ship dynamic interface, under the aegis of The Technical Co-operation Programme (TTCP), heralded the development of generic ship configurations, namely, Simplified Frigate Ships (SFS1 and SFS2) to validate various CFD codes and to generate validation data through model testing. The goal is to progress the application of simulation to maritime helicopter design, test and evaluation with increased pace and effectiveness through collaboration. The high potential payoff from shared learning has proved to be a strong motivation for the different nations to align activities where possible and to present and discuss the various facets of their work openly with other TTCP members [33] Various other research groups outside the TTCP have shown interest to study the SFS. For instance, Praveen et al. [34] and Makkar et al. [35], have also conducted research on the SFS models and published their findings. There is no aircraft carrier equivalent of the simplified frigate ship at present. This paper conceptualizes such a generic aircraft carrier.

A generic carrier must be a typical and representative case of study for flow over the flight deck of aircraft carriers. It must have a simplified geometry for ease of manufacture, replication in experimental studies and meshing requirements of computational simulations. It must manifest the common and most relevant aerodynamic aspects of typical aircraft carriers. The salient features that are characteristic of carriers which must be infused into a generic carrier are listed below.

- Asymmetry of the flight deck about centre line.
- Angled landing strip for simultaneous launch and recovery.
- Flared out hull rising above the water.
- Flight deck beam much wider than the beam at the water line.
- Large overhangs of deck with thick sponsons attached.
- Sharp edges and 'notches' on the flight deck.
- Island superstructure typically of the size of about 2-3% of gross volume of ship, placed aft of midship, along the starboard side.

We studied several different aircraft carriers in service to understand their key common attributes. We have derived a simplified carrier geometry, the Generic Aircraft Carrier from these studies. The dimensions and details of the GAC geometry is illustrated in Figure 1. The GAC inherits relevant traits from most carriers, while still remaining simplified enough to fabricate and model numerically.

7. EXPERIMENTAL SETUP

This section describes the experimental procedure undertaken as part of this study. The model fabrication and installation are described in detail. The instrumentation, apparatus and conduct of experiments is explained.

7.1 GENERIC AIRCRAFT CARRIER MODEL

A wooden model of the Generic Aircraft Carrier described in the previous section was fabricated for conducting tests in the wind tunnel, shown in Figure 2. The assembled model consists of three parts, namely, the hull and the flight deck fused together, and a removable island superstructure as shown in Figure 3. The current experiment aims to capture the pressure distribution over the deck of the carrier. Towards this objective, a total number of 105 pressure taps were inserted into the flight deck by carefully drilling holes at various locations. The siting of these taps was done to capture data at as many discrete points as possible subjected to the practical restrictions imposed by the fabricated model such as joints, corners, etc. The locations of the pressure taps are shown in Figure 4. The metallic pressure taps were made to sit flush with the surface of the flight deck to ensure that that they do not interfere with the flow. Plastic tubing leads of sufficient length were drawn out from these pressure taps inside the hull of the carrier and systematically labelled with identification tags. The assembled model with the pressure taps and tubing leads is shown in Figure 5. The flight deck that extends well beyond the breadth of the hull in aircraft carriers, presented a challenge in installing the pressure taps. The flight deck component was fabricated as a hollow structure to accommodate this installation. The plastic tubing had to be adeptly manipulated to remain within the overhang portion and led through channels drilled into the junction of the hull and flight deck to group with the other tubing below ship's centreline within the hollow hull. The upturned model with exposed tubing leads is shown in Figure 6.

7.2 WIND TUNNEL AND INSTRUMENTATION DETAILS

The experiments are conducted in an open circuit type blow-down wind tunnel consisting of two diffusers, settling chamber, contraction cone, and a test section of 0.75 m x 0.45 m cross-section having a length of 5.9 m available at IIT Delhi. The settling chamber is provided with a honeycomb and screens to ensure a uniform flow in the test section. The contraction cone has a contraction ratio of 12.8 in two stages. A blower driven by a 30 HP motor through a belt drive provides the airflow. The maximum velocity attainable in the wind tunnel is 24 m/s. Different air velocities can be attained by controlling the speed of the blower fan. A standard Pitot-static tube is inserted into the wind tunnel, upstream of the model. The readings of static and dynamic pressures measured by the Pitot-static tube were recorded for estimation of free stream velocity accurately. For the present case, the free stream velocity was maintained at 13.72 m/s. This corresponds to a Reynolds number of 1.062×10^6 based on model length as the characteristic length scale.

7.3 MODEL INSTALLATION AND EXPERIMENTAL PROCEDURE

These experiments are aimed at generating baseline pressure distribution over the 'flat' flight deck of GAC, with the island removed. The model was mounted on the base of the wind tunnel test cross-section as shown in Figure 7 and bolted in place. Plexiglas windows were installed to clearly view and photograph the model during the experiments. The orientation of the model is such that the aircraft landing line aligns with the wind tunnel centre line. This is in accordance with standard carrier operating procedures for aircraft recovery to have wind flow down the angled landing strip. The ship's centre line is hence, at a 9-degree angle, starboard of the centreline. The slanted aft deck edge of the model is therefore perpendicular to the wind tunnel centre line. Table 1 describes the position of the model in the wind tunnel.

Cross section	0.75m X 0.45m
Inlet	2.18 forward of model fore
Outflow	2.62m aft of model aft
Blockage	5.15%

Table 1. Position of Model in Wind Tunnel

The tubing leads were drawn out from a central opening made on the wind tunnel base below the mounted hull. These leads were connected to the arms of three multitube manometer banks with 36 channels each. One arm in each manometer bank is kept open to the atmosphere to act as a reference. The manometer banks consist of a height adjustable well type reservoir with the water column arms inclinable to improve the sensitivity of the measurements. The experiment was conducted at an inclination of 60 degrees to the vertical. The readings were suitably corrected later to calculate the vertical height of the water column recorded in each arm. The manometer readings (in mm H_2O) were converted to obtain the local static pressure (in Pascals) at each location using the measured free stream pressure in the tunnel and ambient atmospheric pressure as reference. The pressure distribution thus obtained over the flight deck is shown graphically in Figure 8.

8. NUMERICAL MODELLING

CFD simulations were undertaken using commercial software package Ansys FLUENT®. The geometry of the model was numerically modelled, and domain was set up to replicate the experimental setup. The origin was fixed at the centre of the carrier's transom deck edge, with X axis pointing downstream, aligned with the wind tunnel centre line (and hence aircraft landing line). The Y axis is oriented starboard and lies on the transom edge. Mutually perpendicular Z axis points upwards. The boundary conditions setup were velocity inlet and pressure outlet with no-slip condition imposed on wind tunnel walls and the body surface.

The meshing employed was structured hexa type and initially the domain consisted of about 4 million cells. This was further adapted to have a finer mesh near the wall (flight deck) thus increasing the mesh size to about 7.5 million elements. Since the objective of this simulation, was an unconstrained exploration to find the most suitable turbulence model, several turbulence models from the software's library were used. CFD runs were undertaken with the one-equation Spalart- Almarras model. Popular two equation models k-ω, k-ε models and their variations (including EWT-Enhanced Wall Treatment), were also employed. Another two-equation option involving laminar kinetic energy, k-kL turbulence model was explored. Also tested, was the four-equation transition SST model. Another family of turbulence models available was the elaborate Reynolds stress models (RSM). Reynolds Stress Models closes the Reynolds-Averaged Navier-Stokes equations by solving additional transport equations for the six independent Reynolds stresses. Four variations of RSM models were assessed. In total, thirteen turbulence models were evaluated to find the most adequate model for this problem.

9. RESULTS AND DISCUSSIONS

The pressure distribution measured experimentally (Figure 8) is compared with the pressure contours obtained from numerical simulations. Figure 9 shows the results of CFD analysis obtained using the k-ω SST turbulence model. Qualitatively, the CFD results exhibit good agreement with the experimental plot.

For validation of the numerical model and to choose the most suitable turbulence model that best represents the physics of this problem, a quantitative analysis was carried out. The local pressure values were extracted from each of the pressure tap locations from the numerical model. Pressure coefficient (C_p) at i^{th} pressure tap location, defined as given in Equation 1, was calculated at each location.

$$C_{p_i} = \frac{p_i - p_\infty}{\frac{1}{2}\rho V^2} \quad (1)$$

This was compared against the pressure coefficients obtained from the corresponding pressure tap experimentally. An error index E, defined in Equation 2, was computed to obtain a global measure of agreement that a turbulence model offers with experimental data. It's a measure of the average absolute error in CFD prediction against experimental data, expressed as a percentage.

$$E = \frac{1}{N}\sum_{i=1}^{N}\sqrt{\left(\frac{C_{p_{i\,(cfd)}} - C_{p_{i\,(exp)}}}{C_{p_{i\,(exp)}}}\right)^2} \quad (2)$$

The exercise is repeated with results obtained from simulations using all 13 turbulence models employed in this study. A comparison of the error indices obtained in each case is given in Table 2.

Turbulence Model	Error (%)
k ω - BSL (Baseline Turbulence Model)	14.77 %
k ω - SST (Shear Stress Transport)	7.69 %
k ε- RNG Scalable WF (Scalable Wall Function)	27.08 %
k ε - RNG EWT (Renormalisation Group EWT)	14.61 %
k ε - Realizable EWT (Enhanced Wall Treatment)	13.63 %
k ε - Realizable Scalable WF (Scalable Wall Function)	30.96 %
Transition SST (4 equation Shear Stress Transport)	8.94 %
k-kL transition	14.04 %
Spalart- Almarras	14.93 %
RSM Stress Omega (Reynolds Stress Model)	16.94 %
RSM Stress Omega (Linear Reynolds Stress)	14.56 %
RSM Stress Omega (Non-Linear Reynolds Stress)	24.35 %
RSM Stress BSL (RSM Stress Baseline)	14.13 %

Table 2. Comparison of Prediction Errors by Different Turbulence Models

From the analysis, it emerges that the SST k-ω model shows the best agreement with experimental data with a prediction error of 7.38%. The shear-stress transport (SST) k-ω model was developed to effectively blend the robust and accurate formulation of the k-ω model in the near-wall region with the free-stream independence of the k-ε model in the far field. The four-equation transition SST emerges as the second-best turbulence model with 8.94% error. The Transition SST model is based on the coupling of the SST transport equations with two other transport equations, one for the intermittency and the other for the transition onset criteria, in terms of momentum-thickness Reynolds number.

To appreciate the pressure variation along the model geometry, the pressure coefficient, C_p obtained from experiments and numerical simulations were plotted along multiple 'lines' on the body geometry. A comparison of experimental data and CFD data obtained for all turbulence models is plotted along the ship's centre line in Figure 10. Figures 11, 12, 13, 14 and 15 depict the variation of C_p obtained from numerical simulations using SST k-ω along the centre, port, starboard, outer port and outer starboard lines on the body geometry. Each plot has an inset depicting the line of pressure taps being analysed.

Analysis of the pressure measurements over the deck reveals strong regions of low pressure on the bow and forward port shoulder. The hull being at an angle to the flow, the effects of this low-pressure region seem to travel downstream with the flow. The interaction of the separated flow originating from this region, with the flow further downstream, may be responsible for the pressure distribution over the outer starboard line exhibiting low magnitudes and a wavy nature. To investigate these effects further, velocity contours across a transverse plane erected at the ship's stern were plotted and are shown in Figure 16. Velocity deficits of about 20%, compared to the free stream, are observed in this region. These deficits are what pilots perceive as the 'burble' further downstream in the flight approach path.

10. CONCLUSIONS

This study has underlined the requirement of incorporating air wake studies as a design driver in the design of future aircraft carriers. The burble behind the carrier and the problems thereof, have been discussed. The paper has proposed a systematic research procedure to confront the problem in a more generic and fundamental manner. A simplified carrier model, namely, the Generic Aircraft Carrier (GAC) has been conceptualized as a benchmark model to study the problem. Baseline experimental data has been generated by wind tunnel measurements using a wooden model of GAC. The understanding of flow physics obtained by analysing the deck pressure distribution aids in explanation of the characteristic velocity deficits in the 'burble' region. Multiple CFD simulations have been carried out to choose the best turbulence model. The model that offers best validation results with the experimental data for this problem is the SST k ω model. Further experiments are in progress to expand the baseline database to the burble region and investigate the effects of island placement and configuration.

11. ACKNOWLEDGEMENTS

We thank Cdr B Praveen and Cdr IS Makkar, pursuing research in the area of ship wake and helicopter interactions at IIT Delhi for providing insight and expertise that greatly assisted the research. We thank Lt Naresh Kumar for his enthusiastic and valuable support in the experimental setup.

12. REFERENCES

[1] LEHMAN A.F, 'An Experimental Study of Dynamic and Steady State Flow Disturbances Encountered by Aircraft During Carrier Landing Approach', *Office of Naval Research Report #64-16, Oceanics Incorporated*; September,1964

[2] HOIBERG, A; 'Assessing The Health Risks Of Carrier Landings In Us / Navy Pilots' *Naval Health Research Center San Diego,* 1985

[3] BRICTSON, C. A.; PITRELLA, F. B; WULFECK, J. W; 'Analysis of Aircraft Carrier Landing Accidents (1965-1969)', *Dunlap and Associates, Washington, D.C.*, November 1969

[4] GOORNEY, A. B; 'The Human Factor in Aircraft Accidents- Investigarion of Background Factors of Pilot Error Accidents' *Flying Personnel Research Committee, Minsitry of Defense, Great Britain*, July 1965.

[5] COLLICOT, C. R; MUIR, D. E; 'U S Naval Aviation Safety', *OEG Study 766*, October 1972.

[6] DURAND, T. S; TEPER, G. L; 'An Analysis of Terminal Flight Path Control in Carrier Landing', *Systems Technology Inc. Technical Report No 137-1*, 1964

[7] OLENSKI, J, J; DANKWORTH, T. P, 'Carrier Crosswind Launching and Landing Operations with Current Navy Airplanes', *Ft 35-093,Navy Air Test Center, Patuxent River*, August 1958.

[8] KJELLMAN, T, K; COLT, R, B; 'Survey Report of Wind Direction and Speed Across Flight Deck USS Ranger CVA 61(CVA 50 Class)' *Index NS 681-026-03, Contract No 72137,Bendix Aviation Corporation, Frietz Instrument Division, Baltimore*, April 1958.

[9] HANNEGAN, E, A; BADGER, H, J; 'Optimum Wind Over Deck for Shipboard Recovery Operations with Carrier Based Planes'; *PTR-RSSH-31003, Naval Air Test Center, Patuxent River, Maryland;1 February* 1961.

[10] OLDMIXON, LIEUT, W. J, 'Acquisition, Reduction and Analysis of Turbulence Data Associated with PA Configuration Approaches to Carrier Landings', *Princeton, University Report #653, Dept of Aeronautical Engineering, Princeton*, July 1963

[11] 'Calibration Report of Wind Direction and Speed Equipment Aboard U80 Kitty Hawk CVA 63', Bendix Aviation Corporation, *Frietz Instrument Division*; August 1962.

[12] 'Survey Report of Wind Direction and Speed Across Flight Deck go USS Enterprise CVA(N) 65' *Bendix Aviation Corporation, Frietz Instrument Division*, March 1962.

[13] BARNETT, W, F; WHITE, H, E; 'Comparison of the Airflow Characteristics of Several Aircraft Carriers', *David Taylor Model Basin*, 1963.

[14] RAJAGOPALAN, G; SCHALLER, D; WADCOCK, A, J; YAMAUCHI, G, K; HEINECK, J, T; SILVA, M, J; 'Experimental and Computational Simulation of a Model Ship in a Wind Tunnel'. *AIAA Paper No. 2005-1347, Presented at the AIAA 43rd AIAA Aerospace Sciences Meeting and Exhibit, Reno, NV*, Jan. 10-13, 2005.

[15] CHERRY, B, E; CONSTANTINO, M, M; 'The Burble Effect: Superstructure and Flight Deck Effects on Carrier Air Wake'; *United States Naval Academy, Annapolis, MD, 21412*, 2010.

[16] MOORHOUSE, D, J; WOODCOCK, R, J; 'Background Information and User Guide for MIL 8785-C', *Military Specification-Flying Qualities of Piloted Airplanes*, 1982.

[17] STEINBERG, M; 'A Fuzzy Logic based Aircraft Landing System', *Lehigh University*, 1991.

[18] 'Department Of Defense Handbook- Flying Qualities Of Piloted Aircraft', *MIL-HDBK-1797 19*, December 1997.

[19] RUDOWSKY, T; 'Review Of The Carrier Approach Criteria For Carrier-Based Aircraft Phase I'; *Final Report NAWCADPAX/TR-2002/71*, Oct 2002.

[20] POLSKY, S; BRUNER, C, W, S; 'A Computational Study of Unsteady Ship Airwake', *RTO AVT Symposium on Advanced Flow Management: Part A – Vortex Flows and High Angle of Attack for Military Vehicles, Loen, Norway*, 7-11 May 2001.

[21] POLSKY S; NAYLOR; 'CVN Airwake Modeling and Integration: Initial Steps in the Creation and Implementation of a Virtual Burble for F-18 Carrier Landing Simulations', *AIAA Modeling and Simulation Technologies Conference and Exhibit*, 2005.

[22] SHIPMAN, J; ARUNAJATESAN, S; MENCHINI, C; SINHA, N; 'Ship Airwake Sensitivities To Modeling Parameters', 2005.

[23] KELLY, M; WHITE, M; OWEN, I; STEVEN, H, J; 'The Queen Elizabeth Class Aircraft Carriers: Airwake Modelling and Validation for ASTOVL Flight Simulation' 2016.

[24] KELLY, M; WHITE, M; OWEN, I; STEVEN, H, J; 'Using airwake simulation to inform flight trials for the Queen Elizabeth Class Carrier' 2016.

[25] MEIER, H, A; GALE, P, A; 'Alternatives in Aircraft Carrier Design', *Naval Engineers Journal, American Society of Naval Engineers*, 1977

[26] RINGLEB, F; 'Airflow Control Arrangements over Aircraft Carriers', *US Patent 3,362,368*, January 9, 1968.

[27] PHILIP, A; 'Aircraft Carriers and Methods of Operation Therof', *US Patent No US6279500 B1*, Aug 28, 2001

[28] NANGIA; R, K; LUMSDEN, R. B; 'Novel Vortex Flow Devices - Columnar Vortex Generators Studies For Airwakes', *34th AIAA Fluid Dynamics Conference and Exhibit 28 June - 1* July 2004

[29] LAMAR, J, E; 'Modifying Ship Air-Wake Vortices for Aircraft Operations'; *Langley Research Center, Hampton, Virginia, NASA Tech Briefs*, 2004Z

[30] LANDMAN, D; LAMAR, J, E; SWIFT, R; 'Particle Image Velocimetry Measurements to Evaluate the Effectiveness of Deck-Edge Columnar Vortex Generators on Aircraft Carriers', *RTO-MP-AVT-124 , NATO-PFP Unclassified*, 2005

[31] CZERWIEC; R, M; POLSKY, S; 'LHA Airwake Wind Tunnel and CFD Comparison with and without Bow Flap', *22nd Applied Aerodynamics Conference and Exhibit 16 - 19* August 2004

[32] MUNGUÍA, B, C; BUI, N, V; JAMES; LEWI, J, B; RICHIE, D, R; 'Aircraft Carrier Burble Mitigation With Alternating Current Dielectric Barrier Discharge Plasma Actuators', *53rd AIAA Aerospace Sciences Meeting* 2015

[33] WILKINSON, C, H; ZAN, S, J; GILBERT, N. E; FUNK, J, D; 'Modeling and Simulation of Ship Air Wakes for Helicopter Operations- A Collaborative Venture', *Paper No. 8, Proceedings of NATO RTO Meeting on Fluid dynamics Problems of Vehicles Operating Near or in the Air-Sea Interface, RTO-MP-15,*

Neuilly-Sur- Seine Cedex, France, February 1999.

[34] PRAVEEN, B; VIJAYAKUMAR, R; SINGH, S, N; SESHADRI, V; 'Parametric Investigation of Airflow Over Helodeck of a Generic Frigate with Introduction of Flow Modifying Device' *International Conference on Computational and Experimental Marine Hydrodynamics MARHY* 2016

[35] MAKKAR, I, S; VIJAYAKUMAR, R; SINGH, S, N; SINHA, S, S; 'Study Of Dynamic Flow Effects Due To Ship Air Wake And Rotor Downwash Interaction On Warship Helo Decks', *The Royal Institution of Naval Architects International Conference Warships 2016: Advanced Technologies in Naval Design, Construction & Operation*, 2016.

13. AUTHORS' BIOGRAPHY

Lt Cdr K Vignesh Kumar is a graduate of Naval Architecture from the Cochin University of Science and Technology. He is a post graduate in Naval Construction from Indian Institute of Technology, Delhi and M Tech in Ocean Engineering from Indian Institute of Technology, Kharagpur. He is presently appointed as Adjunct Faculty at the Naval Construction Wing, Department of Applied Mechanics at Indian Institute of Technology, Delhi. The study was undertaken as part of his doctoral research work at IIT Delhi. He is a member of the Institution of Naval Architects, India. His research interests include Ship Hydrodynamics, Warship Design and Aircraft Carrier concepts.

Cdr MP Mathew is a graduate of Naval Architecture from the Cochin University of Science and Technology. He is a post graduate in Naval Construction from Indian Institute of Technology, Delhi and M Tech in Ocean Engineering from Indian Institute of Technology. He is presently the Officer in Charge of Naval Construction Wing, IIT Delhi. His research interests include ship dynamics and computational fluid dynamics.

Prof. SN Singh is an Aeronautical Engineer. He did his B.Tech. as well as M.Tech. in Aeronautical Engineering from Indian Institute of Technology, Kanpur and Ph.D. in Fluid Engineering from Indian Institute of Technology, Delhi. He is presently Professor in the Department of Applied Mechanics, Indian Institute of Technology, Delhi. He has over 37 years' experience in teaching and R&D. He has published 300 research papers. He has so far guided 25 Ph.D. Projects and 100 M.Tech. Projects. He is a consultant to a number of industrial organizations and has so far completed more than 1000 consultancy and sponsored projects in association with Prof. V Seshadri. His areas of interest are Fluid Mechanics, Internal Flows, Computational Fluid Dynamics, Two Phase Flows and Flow Instrumentation. He is a Life member of AIAA, Fellow of the Institute of Engineers, Life Member of National Society of Fluid Mechanics and Fluid Power (NSFMFP), Indian Society of Mechanical Engineers (ISME), Indian Society of Technical Education (ISTE) and Aeronautical Society of India.

Dr Sawan S Sinha is an Aeronautical Engineer. He did his B.Tech. in Aerospace Engineering from Indian Institute of Technology Kanpur and PhD in Aerospace Engineering from Texas A&M University. He has worked with Defence Research & Development Organization (DRDO), Hyderabad and EACoE, John F. Welch Technology Centre (GE India), Bangalore. Post PhD he worked as Research Associate at Aerospace Engineering Department, Texas A&M University. He has published 19 research papers in national & international refereed journals/conferences. His research interests include Theory and Modelling of Compressible Turbulence, Hypersonic Flows, CFD of massively separated flow, Airship design, Flow-thermodynamics of steam turbines at low-load condition.

Dr R Vijayakumar is a graduate of Naval Architecture from the Cochin University of Science and Technology. He has done a PG Diploma in Naval Construction from Indian Institute of Technology, Delhi, M Tech in Ocean Engineering from Indian Institute of Technology, Madras and Ph.D. from Indian Institute of Technology, Delhi. He had served in Indian Navy for last 21 years, in dockyards and design directorate. He is presently working as assistant professor at the Ocean Engineering Department, at Indian Institute of Technology, Madras. Where he teaches the subjects of Ship Hydrodynamics, Ship Dynamics and Advance marine vehicles. He has published a total of 25 technical papers in referred Proceedings of National Conferences. He is a member of the Institution of Naval Architects, India. His fields of interest include Warship Design, Ship Hydrodynamics, Ship Dynamics and Computational Fluid Dynamics (CFD).

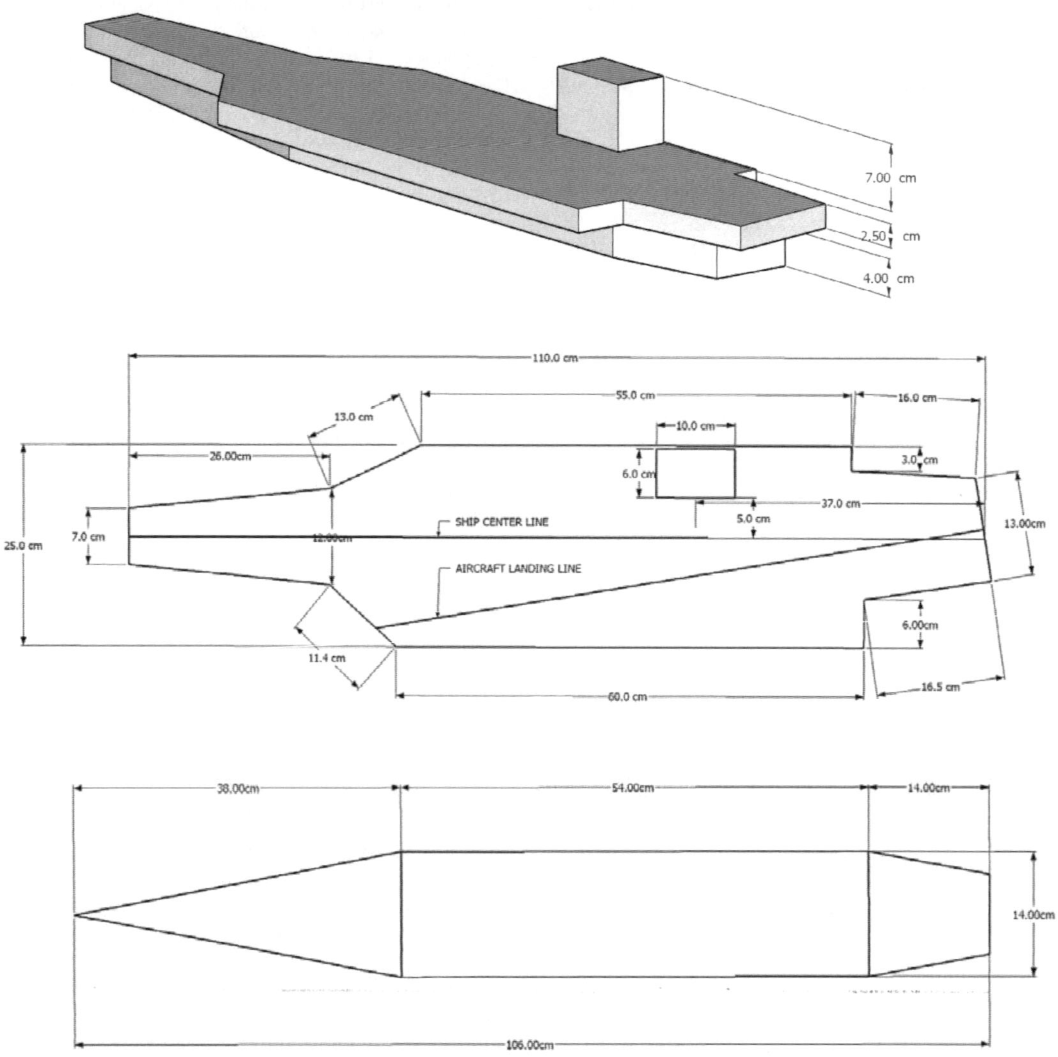

Figure 1: Details of the Generic Aircraft Carrier Model.

Figure 2: Wind Tunnel at IIT Delhi.

Figure 3: Photograph of the assembled GAC model.

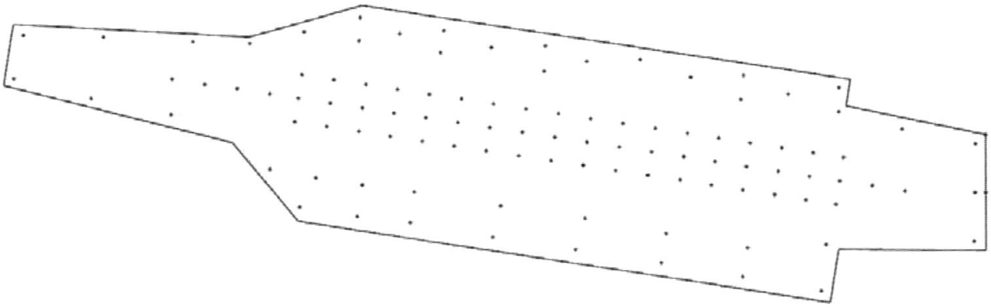

Figure 4: Position of Pressure Taps on the Model.

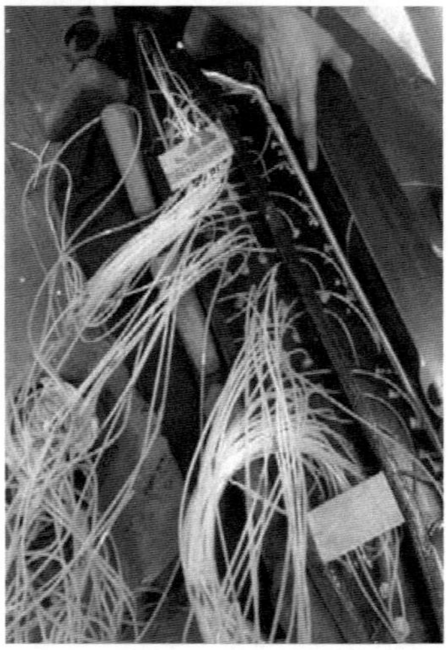

Figure 5: Model with details of deck overhang tubes.

Figure 6: Model with exposed tubing

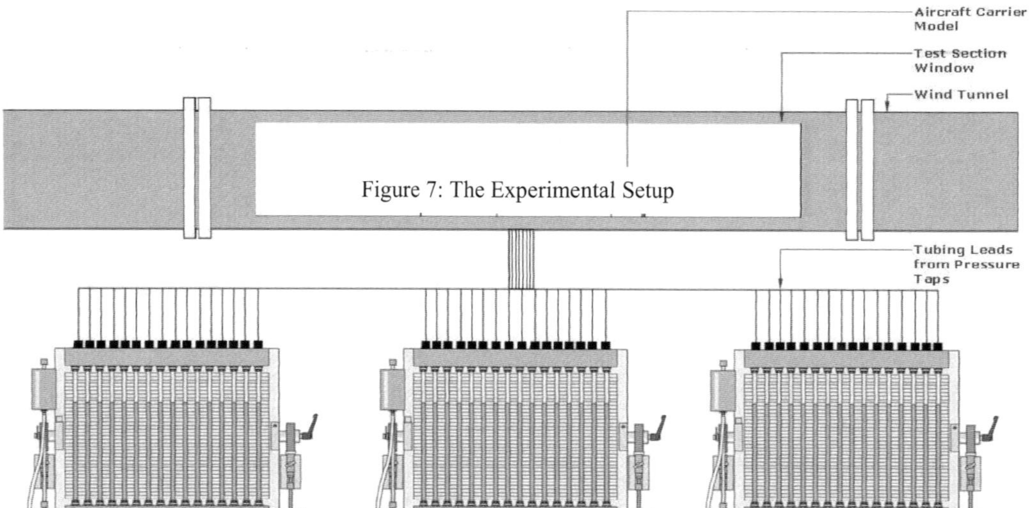

Figure 7: The Experimental Setup

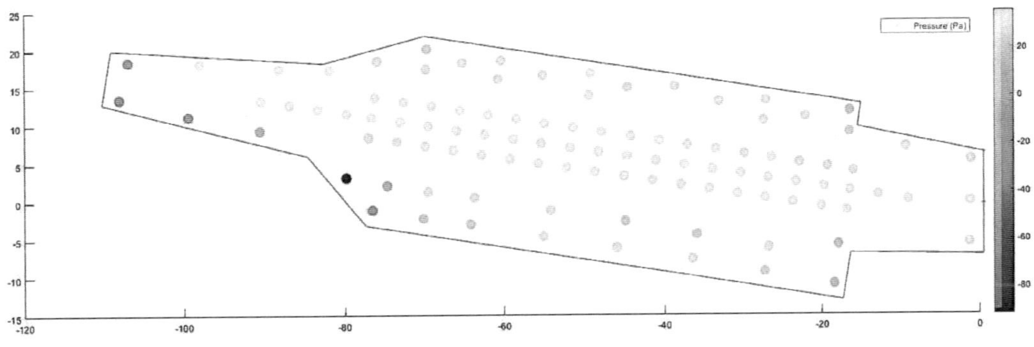

Figure 8: Results of Experiments – Static Pressure Measured at Pressure Taps

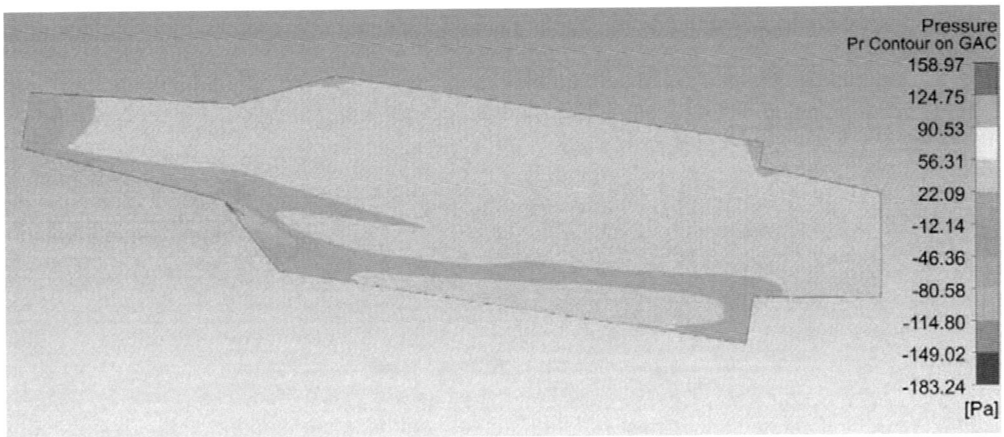

Figure 9: Results of CFD Simulation – Static Pressure Contours on Deck using SST k ω model.

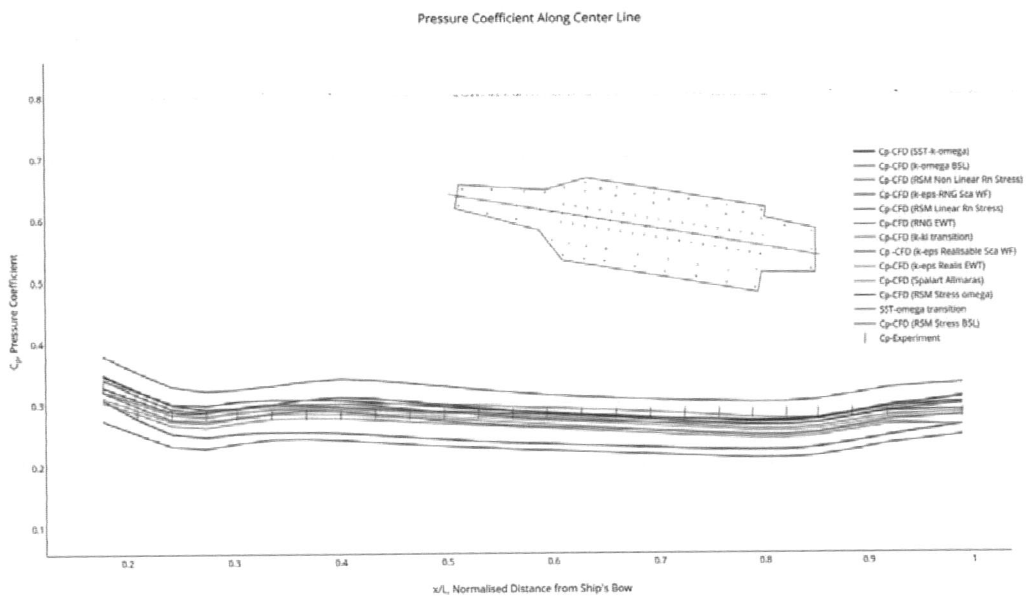

Figure 10: Comparison of Cp along Centre Line – All Models with Experimental Data

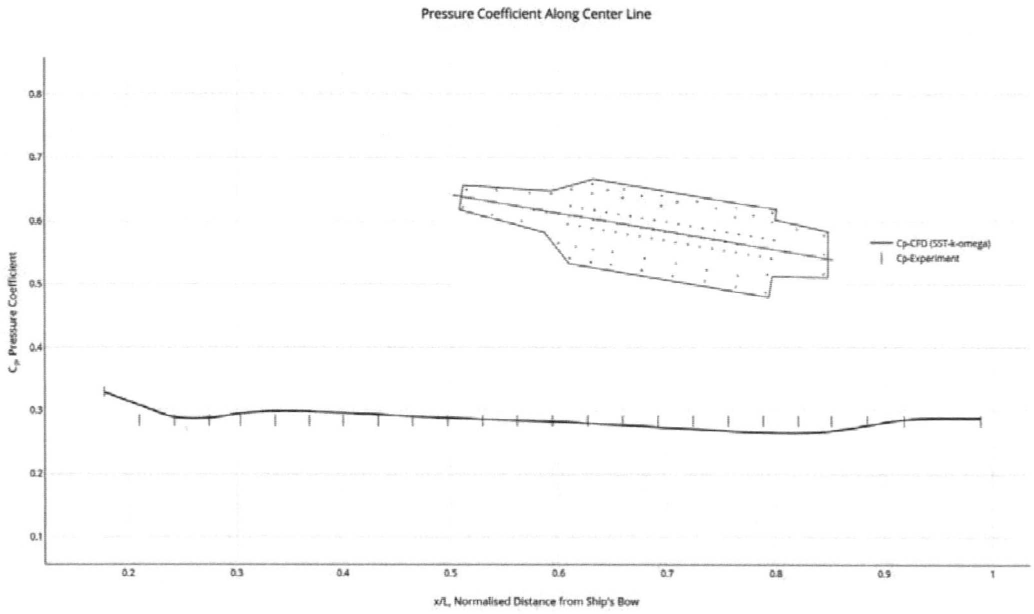

Figure 11: Comparison of Cp along Centre Line – SST k ω Model with Experimental Data

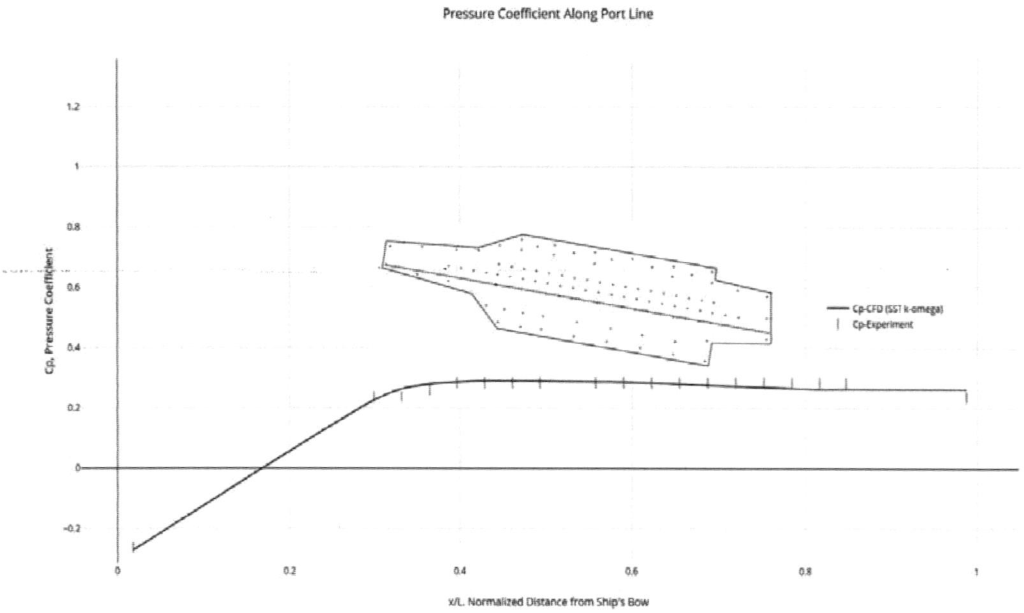

Figure 12: Comparison of Cp along Port Line – SST k ω Model with Experimental Data

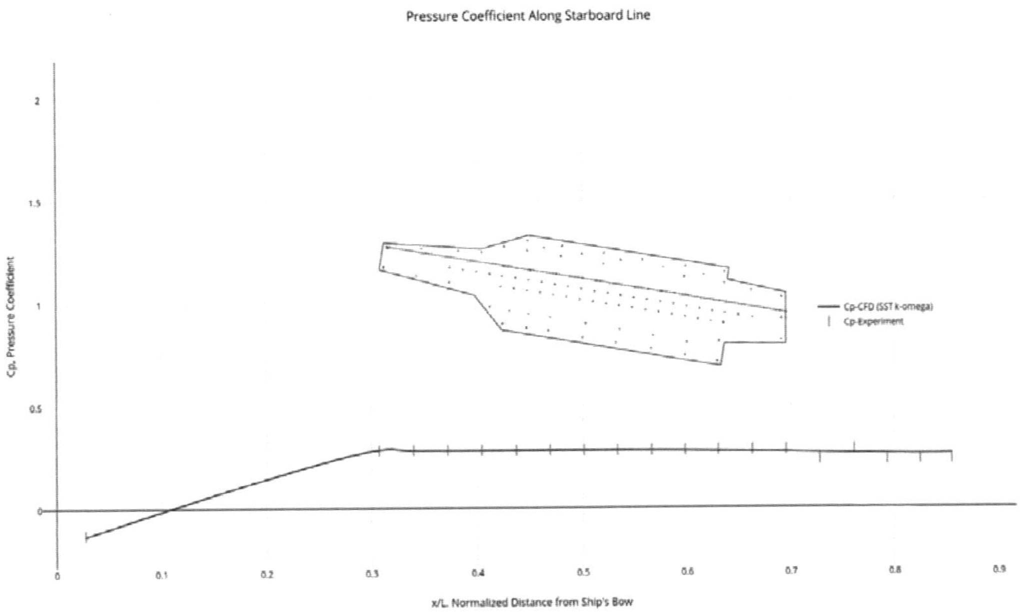

Figure 13: Comparison of Cp along Port Line – SST k ω Model with Experimental Data

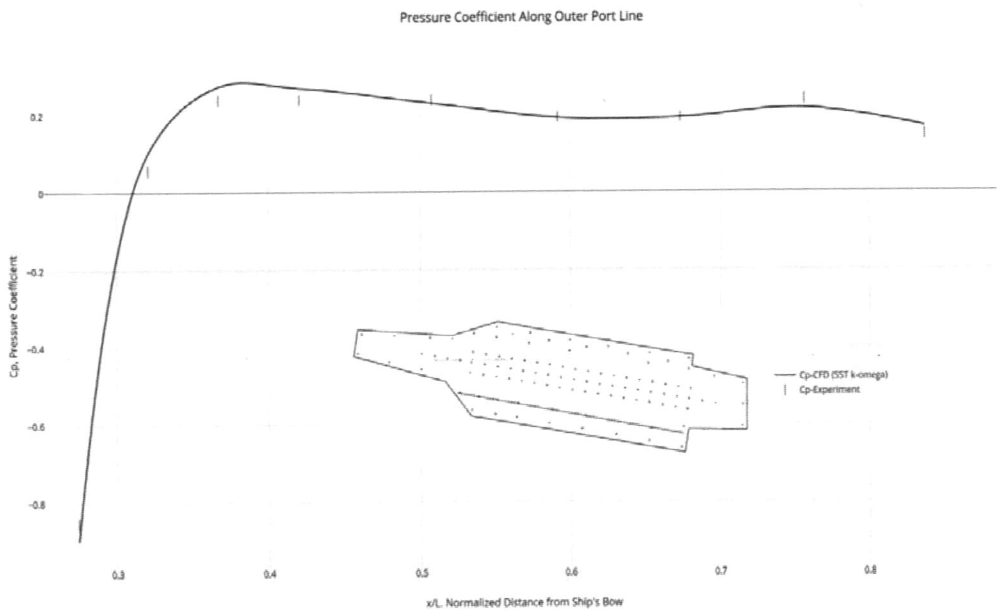

Figure 14: Comparison of Cp along Outer Port Line – SST k ω Model with Experimental

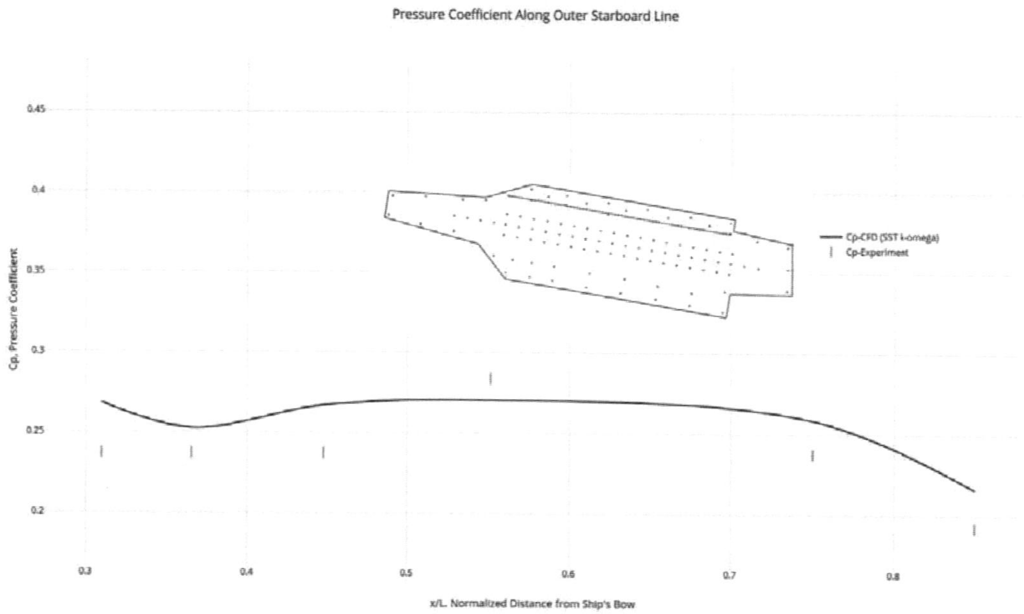

Figure 15: Comparison of Cp along Outer Starboard Line – SST k ω Model with Experimental Data

Figure 16: Velocity Contours along Transverse Plane at the Stern of the Carrier

THE NATO DRIVE TO MISSION MODULARITY

David Manley FRINA, RCNC, UK Ministry of Defence

SUMMARY

The drive to deliver flexible, adaptable ships had caused several NATO navies to consider modular systems as a means of delivering capability. At the same time there is a growing desire to increase scope for interoperability. And the rapid advances in unmanned surface, air and subsurface systems requires a method of delivery of those systems into ships that allows new technology to be embarked quickly without costly modifications to the host platform being required. Because of these imperatives NATO established the Specialist Team on Mission Modularity (ST/MM), working under the NATO Naval Armament Group to deliver standards and guidance in ship and module design, operability, logistics, maintenance and manning. This paper covers the work of ST/MM and its supporting teams within navies and Industry, progress towards the publication of standards and guidance, tests, trials and other activities that have been conducted to inform the programme. It considers the use of 20' ISO containers as a basis for a modular system, looking at a number of enhancements over existing standards to improve their ability to be used in a naval environment whilst still maintaining the essential geometric standards that allow them to seamlessly utilise the international ISO container logistics environment. Finally, it reviews activities addressing impact on ship design and operation, maintenance, logistics and training, before concluding with a discussion of the future NATO programme.

1. INTRODUCTION

Navies are continually challenged by the increasing cost of warship design, manufacture and operation, whilst facing budgetary constraints, pressures on available hull numbers and an increasing range of missions from humanitarian assistance through to all-out war [ref i]. This complex procurement and operational environment is driving navies to actively consider options to increase adaptability and flexibility[1] within their fleets to maintain relevancy for the most time over the duration of a ship's life. A prime enabler in these efforts is Mission Modularity (MM).

Mission Modularity is often confused with modular construction, but it is quite distinct. Mission Modularity seeks to deliver maritime capability into warships and auxiliaries through the use of equipments and systems contained in Mission Modules that are embarked in ships in dedicated upper deck stowage areas or internal Mission Bays. The approach offers a significant range of benefits, not least the potential to rapidly insert new capability, to re-role vessels that can accept modules to cover a range of tasks and to allow ships to be outfitted for effective operation in niche, limited duration tasks. Earlier studies for the UK MOD [Ref ii] demonstrated the benefit of a modular force over a traditional force[2] in enabling a fleet to meet a diverse range of standing and emergent tasks. The study, conducted in support of early Type 26 concepts, analysed the degree to which traditional and modular-based forces achieve a range of standing tasks typical of those conducted by the RN and emergent tasks in UK waters and around the world, assessing the shortfall in being able to meet those tasks. The study showed that the modular force achieved a significantly higher degree of performance (Figure 1). Benefits and potential penalties are discussed further later in this paper.

2. DELIVERING MODULARITY

There are a range of delivery methods available, some of which are well established but which do not immediately come to mind in the modularity context. A recent Dstl study [iii] suggested ten broad categories:

- Container (TEU)
- Vertical Launch Missile Silo (e.g Mk. 41, SYLVER) – wide selection of weapons able to be used
- Deck + Compartment Under (StanFlex, US Gun Mission Module - GMM)
- Deck fitting (e.g. Phalanx, Small Calibre Guns)
- USV/UUV + Launch and Recovery System (boat, towed body on ISO skid)
- Large boat/USV/UUV (modular chocks and lifting points, payload craft compatibility with recovery sled, etc.)
- Flight deck + hangar (or vehicle deck)
- Small modules (e.g. Minicons)
- Compartment / Rack / Component modules

[1] "Adaptability" and "flexibility" are terms that are outwardly similar and often confused. In general, maritime parlance "flexibility" refers to the ability of a complete entity, in this case a vessel, to perform more than one function without major change; adaptability is a measure of the ease with which a ship or system design can be changed to produce variants optimised to perform additional or different functions. Mission Modularity can be seen to increase the inherent adaptability in a warship, although potentially at the expense in short term flexibility if the degree to which platform capability is modularised is taken to the extreme.

[2] The paper referred to here uses the term "flexibility" to include modularity

- Personnel detachments (austere accommodation or built in accommodation margin)

There are clearly areas where a modular approach gives rise to significant operational, cost and maintenance benefits. But, as with any "new" concept, there is a risk that Mission Modularity could be taken to be a universally applied panacea without due consideration of the potential drawbacks that could arise. Hence, as with other "novel" concepts a measured approach is needed to determine and assess the benefits and drawbacks of the application of modularity to a specific capability. Work carried out in in the UK and in NATO specialist teams has highlighted a number of potential benefits but also a number of challenges. These include:

Potential Benefits

- **Increased adaptability**; niche capabilities required only for short duration missions can be delivered effectively at a fleet level across a wide range of platforms;
- **Rapid Fit MTE**; the fitting of Mission Task Equipment (MTE) has been described as "tortuous"; the use of mission modularity could significantly enhance the ease with which MTE could be fitted and moved around ships of the fleet;
- **Reduced platform and/or class numbers**; a given fleetwide range of capabilities may be delivered through fewer platforms, or with fewer platform types (for example mine-hunting, hydrographic and patrol capabilities that previously needed the distinct ship types to deliver could be provided by a single class of ship;
- **Decreased maintenance burden**; modules can be stored, maintained and updated in a benign environment ashore when not in use;
- **Training**; mission specialists can train on their equipment in a benign synthetic environment when not deployed, as a supplement to more realistic operational training at sea;
- **Reduced equipment holding**; niche capabilities can be delivered through a reduced stock of equipment. Not every vessel is required to carry all equipments and systems permanently;
- **Increased interoperability**; the ability for NATO navies and allies to work seamlessly together. This aspect has drawn the attention of the NATO Smart Defence Initiative;
- **Capability Development**; modularity enables a continuous development for a range of capabilities, irrespective of steps in the timelines of replacement classes; in this way there is the potential to de-link 'once in a generation' development opportunities from developments in discrete capabilities. This can de-risk new capabilities, and enables opportunities to exploit emerging technologies far more quickly than in legacy platforms.
- **Existing Transportation Infrastructure**; this is particularly relevant where a modularity solution is delivered using ISO containers; there is a worldwide infrastructure on land, at sea and in the air dedicated to the rapid and effective movement of containers. This infrastructure could facilitate rapid deployment of modules to and from a home base to a forward operating base and then on to ships.

Potential Challenges

- **Design impact**; the need to include large mission bays with ready access to shoreside facilities high in a ship may drive designs that are larger than more traditional designs. A need to include power, water cooling, ventilation, waste removal and other services to modules may drive larger capacities in ships' systems and accommodation if not delivered through modular means;
- **Reduced flexibility**; an imbalance in the degree to which capability is permanently fitted within a vessel and delivered through modular methods may impact on the short-term flexibility traditionally enjoyed by warships;
- **Capability Compromise**; Capabilities delivered through modular means may be compromised to some degree when delivered through a modular approach (for example; can an area air defence capability be effectively fitted into a modular system?). This will either drive an acceptance of a compromised capability or will prevent the application of modularity to some capabilities;
- **Crew integration and teamwork**; personnel in mission specialist teams embarked in ships to support modular capabilities may find it challenging to integrate effectively with standing ships personnel. Embarked mission specialists may also suffer from unfamiliarity with general shipboard routine, and also with platform specific characteristics that may be challenging in an emergency situation.

3. MISSION MODULAR WARSHIP PROJECTS

The Royal Danish Navy has operated vessels using the StanFlex modular system for many years [ref. iv]. Since its adoption in the RDN's *Flyvefisken* class corvettes nine classes of warship and auxiliary have been equipped to accept StanFlex modules, with roles ranging from oceanographic survey and pollution control to anti surface and anti-air missile systems. There are currently over 20 vessels operating the system (see Figure 2). However, StanFlex is very much a national solution that, despite its success, has not been used outside Denmark.

In more recent years the US Navy has fielded modular systems in the two Littoral Combat Ship classes [ref v], with their modular capability based largely on 20' ISO containers and their associated footprint and securing arrangements. These ships also use the US Gun Mission Module system (GMM), similar in some ways to StanFlex in that modules are dropped into dedicated spaces from above.

Less formally, many nations have deployed mission specific equipment in ships based in 20' containers, skids and other rapid fit systems. For example, the Royal Navy (RN) and Royal Fleet Auxiliary (RFA) have deployed container based Unmanned Air Vehicle (UAV) systems, and several RN and RFA vessels are outfitted to carry disaster relief, temporary accommodation, training facilities and other stores in deck mounted container positions.

However, this situation is set to change drastically. Several NATO and allied navies have embarked on surface warship programmes that include Mission Modularity concepts. The UK Type 26 and Australian Hunter class frigates include a midships Mission Bay, capable of hosting up to ten 20' ISO containers, with a number of mission roles currently envisaged. The Mission Bay is equipped with a highly capable handling system that doubles as a boat launch and recovery system, and which also allows the ship to load or unload containers to and from the pierside.

The Italian Pattugliatori Polivalenti d'Altura (PPA) light frigate takes a similar approach but with an exposed midships mission area. This allows the ship to carry up to eight 20' ISO containers (four footprints, double stacked) with a centreline crane for container load/unload and boat launch. A second internal mission bay beneath the flight deck allows the carriage of another five containers.

Elsewhere, the requirements for Germany's MKS-180 multi role frigate include the carriage of modular capability in 20' ISO containers, and the US Future Surface Combatant programme is also likely to include a modular capability requirement.

Mission Modularity as a concept is not restricted to NATO and allied navies. The Russian Project 22160 class corvette features an aft mission bay below the flight deck. This is used to provide extra living space for crew or special operations personnel, launch and recovery through stern doors of small boats and underwater or surface unmanned craft, and for a containerized drone control centre. But the space's primary purpose is to house containerized long-range weapons, with up to eight Klub-N long range anti-ship or land attack cruise missiles being carried as required.

4. THE ROYAL NAVY MODULARITY STRATEGY

Mission Modularity as a concept requires direction if it is to be adopted and drawn into a navy's DNA effectively. In the UK the adoption of the Type 26 mission bay, increasing use of autonomous systems and concepts in mine counter measures to deliver offboard solutions is driving the development of the Royal Navy Modularity Strategy. The Strategy is currently under development by Maritime Capability team (MARCAP) within Navy Command Headquarters (NCHQ), assisted by the Naval Authority Group (NAG) within the MOD Submarine Delivery Agency (SDA). When completed and endorsed it will provide the strategic direction for mission modularity in the RN, directing research and development and providing the foundation for platform and equipment project teams to develop and deliver capability appropriate for modularity in an effective manner. Interoperability with partner nations is a key consideration in the Strategy, hence it recognises the NATO direction in this area and drives towards the adoption of common standards, interfaces, concepts of operation, etc. It is currently intended to publish the Strategy in 2019.

5. DELIVERY OF MODULAR CAPABILTY

Key to the implementation of a modularity strategy is an understanding of the effective scope of modularisation of capability. Whilst it is theoretically possible to apply modularity principles to almost any maritime capability it is important to understand where this can be done practically, both in terms of delivery of a worthwhile capability and impact on the ship. For example, a complex anti air warfare system such as Sea Viper may require an extensive command and control facilities, a complex radar system mounted high in the ship that has very tight alignment tolerances coupled with a large silo. Whilst it would be possible to containerise such a system the extent of the capability in terms of numbers of containers would be significant, alignment of a removable elements could well be an issue, as would topweight issues associated with the radar; as such an AAW system may be better delivered through a permanent fit in a warship. In contrast, a mine countermeasures system could be delivered in a small number of containers comprising a deployment and stowage system for mine detection and disposal vehicles, and a command and control cabin; possibly comprising only three or four containers and able to be embarked in a range of vessels from OPV size or larger, the only significant requirement being access by the deployment module to the sea. The scope to which a warship's capabilities are modularised will also impact on its short-term flexibility; it is important to ensure that a balance is struck, and that those capabilities upon which a warship is frequently required to draw are not affected by an extreme application of a modular philosophy.

Figure 3 gives an indication as to the scope of interface coverage. Physical interfaces (connection of the containers into the ship, provision of power, data and other services) are the most obvious, but often forgotten are broader aspects such as logistics, compatibility of items stored in modules with the storage area (for example explosives or hazardous materials) operational and platform environments, security and electromagnetic interference (for example, TEMPEST considerations and the unwanted impact of electromagnetic emissions from the modular capability on the ship and vice versa). Apparent from this diagram is the plethora of supporting interface documentation that is likely to be generate, including Functional Interface Specifications (FIS), Interface Control Documents (ICDs), and Process Interface Documents (PIDs).

This leads to the critical need to understand the impact on the host vessel of a modular capability. At its most obvious level Mission Modularity requires the vessel to be able to accept containers either on deck or in one or more dedicated mission bays, with space required around modules to be able to access deck and system connections, personnel movement etc. This in itself can impose a significant space requirement, but looking deeper there are questions as to the provision of power and other services (should these be delivered from the ship or contained within additional modules – the former drives additional power and service margin into the ship, the latter eats up mission bay space and may introduce significant safety issues), personnel (are accommodation margins included in the ship, or modular accommodation), and handling (embarked and moved using an integral mission bay handling system or only by shoreside cranes and equipment – see Figure 4).

Modularity impacts the complete spectrum of lines of development[3] and this must be recognised and understood. For example, it is assumed that trained personnel will accompany complex mission module packages such as mine countermeasures or signals intelligence teams; basing these personnel on shore with their equipment whilst not deployed offers attractive benefits in terms of training and harmony requirements, but there are obvious implications for training and experience at sea and integration as part of a ships company. Such issues are not insurmountable and are the subject of further study through NATO and national programmes, but need to be recognised and addressed. And indeed, some aspects may be entirely beneficial; for example, storing mission modules in a land based storage and maintenance facility would reduce the maintenance burden on equipment, and would allow the latest updated and well-maintained modules to be deployed when required.

It can be seen that there are a wide range of design, operational and logistics aspects that need to be understood in order to develop and deploy a modular capability effectively within a Navy, let alone across NATO. It is these aspects that NATO's Specialist Team on Mission Modularity (ST/MM) has been tasked with investigating.

6. NATO'S MODULARITY PROGRAMME

Mission Modularity is currently of significant interest to NATO since it offers the possibility of modular systems owned and operated by one member nation being deployed on the warship of another. For example, a British mine countermeasures team deployed on a French OPV, or an Italian intelligence gathering module on a Dutch frigate. At present, and with a few exceptions, the current applications tend to be nation specific, at least in part. As noted previously, StanFlex enjoys a good reputation within Western navies but has not yet been adopted outside Denmark. Whilst this may be acceptable from a national perspective there are obvious opportunities to increase interoperability and cooperation if a more standardised approach were to be taken. It is in this context that NATO established the Specialist Team on Mission Modularity (ST/MM), supported by a succession of NATO Industrial Advisory Group (NIAG) teams, to investigate and develop international naval standards for Mission Modularity.

ST/MM was established in December 2013 and is one of several teams working under the Ship Design Capability Group (SDCG[4]), itself working to the NATO Naval Armaments Group (NNAG). ST/MM is co-chaired by the United States and the United Kingdom and has an active membership comprising:

- Australia
- Canada
- Finland
- Germany
- Italy
- Netherlands
- Norway
- Turkey
- UK
- US

ST/MM is supported by SDCG's other specialist teams covering Total Ship Systems Engineering (ST/TSSE), Ship Costing (ST/SC), Ship Combat Survivability (ST/SCS), and Seaway Mobility (ST/SM), each of which is tasked with considering Mission Modularity from their individual perspectives. The Specialist Team is just over three years into a multi-year programme aimed at delivering an initial capability in terms of standards and documentation in 2019.

In the short to medium term ST/MM's activities have been focussed on developing interface standards for mission modules based on 20' ISO containers or container skids, along with associated design guidance

[3] In the UK these are referred to as Defence Lines of Development, using the "TEPIDOIL" acronym – Training, Equipment, Personnel, Information, Concepts & Doctrine, Organisation, Infrastructure, Logistics; elsewhere the acronym "DOTMLPF" - Doctrine, Organization, Training, Materiel, Leadership & Education, Personnel, and Facilities is used.

[4] Older readers may remember SDCG as NATO Maritime Capability Group 6

for modules and ships, as well as consideration of wider aspects such as logistics, maintenance and training. Future work is likely to extend into other modular systems.

There is also interest in marinising ground-based modules to supplement the quick localized deployment of humanitarian assistance and disaster relief equipment. Such missions are difficult to design organically in to a balanced ship concept with more stressing mission requirements. Through a NATO standard interface, the concept of deploying mission modules can be realised.

Work is not done within the NATO teams alone. In fact the greater proportion has been and continues to be undertaken in Industry. SDCG has sponsored four NIAG Study Teams looking at aspects of Mission Modularity. SG150 identified the viable scope and initial cost estimates of NATO modularity solutions, and on completion SG168 and latterly SG186 were tasked primarily with the development of module interface standards. Successful completion of this phase of the overall project is essential to the overall mission modularity concept, and the approach taken has been to identify suitable existing standards wherever possible, developing an interface "shopping list" from which a ship or module designer can select the most appropriate types. Over thirty existing ISO and NATO STANAG interface standards have been identified and incorporated, covering the provision of power, fluids and data.

Figure 5 illustrates the scale of the interface issue, and considers only the physical interfaces required for a particular capability (in this case a counter piracy mission package embarked in a typical frigate).

The development of interface specifications only represents a part of the NIAG contribution. These teams also considered wider capability and development aspects, such as transportability[5], storage, personnel management, training and maintenance issues, drawing in expertise and experience form a wide range of industrial partners from across the NATO community. The output from these teams forms the core of ST/MM's developing documentation and standards.

Since standing up in 2013 ST/MM has established its network of supporting teams, developing linked programmes of work with each team to ensure that their specialist knowledge and experience is accessed. ST/TSSE has taken a leading role in considering ship design and operational aspects. Initial studies concentrated on the use of large amphibious warfare vessels to deliver non-warfighting capability (for example humanitarian assistance in the case of a natural disaster). Whilst worthy and credible scenarios they did not present particularly challenging cases given the ease with which containers can be embarked in vessels of this type. Current and future studies are considering more stressing scenarios involving smaller warship types and the embarkation of more warlike payloads, for example mine counter measures systems, signals intelligence modules and close in defensive systems on frigates and offshore patrol ships. Such cases bring with them the challenge of integrating the command and control aspects of the embarked capability with that of the host platform, and spatially challenging requirements for module access to the sky (for example to provide clear firing arcs, access to the electromagnetic spectrum, etc.) or to the sea (for example to launch and recover unmanned systems).

An important element in the scope of mission modularity studies is the cost-benefit analysis (CBA). The objective of the CBA is to give the SDCG credible advice on how useful the concept of mission modularity is to the future NATO maritime forces. The CBA effort is being led jointly by the Specialist Teams on Total Ship System Engineering (ST/TSSE) and Ship Costing (ST/SC), supported by other SDCG specialist teams as required. The analysis is based on a set of naval tasks that must be completed successfully by a maritime task force. Two different forces are being compared: a 'traditional' force without mission modularity (Force A) and a mission modularity focussed force that consists of mission modules and the next generation of ships designed with mission modularity in mind (Force B). In order to produce credible results, the CBA has been split up into two phases. In the first phase, completed in 2017, a small set of naval tasks and a force of only a few assets (ships and mission modules) was considered. Part of this first phase was to learn how operators use Force B differently from Force A. To this end operators, defence planners, and operations planners took part in a tabletop exercise (war game) that was run in Gosport, UK in the summer of 2016 and supported by the UK Naval Design Partnering (NDP) team. The initial point considered a maritime task group engaged in counter piracy operations. Whilst this mission endured the group was, at short notice, required to support a humanitarian search and rescue mission (HSAR), followed shortly thereafter by a non-combatant evacuation operation (NEO) in the region. Each mission required different capabilities at different times within the MTG. During the exercise the participants worked through the changing operational situation while their decisions were recorded, allowing the team to develop a clear understanding of the concept of operations, including for example how assets of a force are assigned to standing and emerging naval tasks.

Phase 2 takes the outcome from the tabletop exercise and uses derived metrics and other data generated by specialist teams to develop a fleet level simulation tool capable of simulating the operation of both forces during several months, years or decades. The simulation tool will be an essential resource for the CBA: a much larger set of naval tasks and many more assets in both forces to be considered in order to produce credible advice on

[5] For example, there is a significant reduction in the selection of transport aircraft that can be used to transport an 8'6" container as opposed to an 8' container; such limitations are not readily apparent to the average naval architect or combat system engineer

mission modularity usefulness. In the second phase Force B will be designed by using the Total Ship System Engineering (TSSE) process: an early stage ship design process based on systems engineering that should enable nations to design more cost-effective warships.

Work to support detailed aspects of mission modularity has also been conducted within navies and defence ministries within NATO which are contributing to the task. The UK has been particularly active, with the following studies focussed on survivability completed or in hand:

- Consideration of the vulnerability of maritime weapon effects on mission modules, and associated mitigations. This has covered underwater shock, blast, ballistic and shaped charge and fragmentation threats. The study also considered basic container ruggedness and, as a by-product, handling on board in the maritime environment. A key outcome from this study was a recommendation to use the DNV 2.7-2 enhanced container manufacture standard developed by Det Norske Veritas (DNV), developed for the offshore industry and able to deliver more rugged containers with improved handling arrangements that does not impact on the basic 20' ISO standard arrangements[vi].

- Development by Thornton Tomasetti Defence Ltd under tasking from the Naval Authority Group of a deck isolation system designed to protect mission modules against the effects of underwater shock. The system provides an arrangement of novel shock mounts located within a rail system that interfaces with the ISO pockets on the container and which can connect to the ship using a variety of methods (e.g. standard "twist-lock" arrangements, bolted pads, etc.). The system has been demonstrated to provide protection to containers and their contents up to shock levels comparable with those for shock hardened or protected equipments in warships.

- Consideration of the deeper integration of a modular capability with the combat system of the host platform. Current studies and concepts have considered the modular capability as somewhat "standalone"" in nature, embarked in the host platform but not connected directly with the host's combat system. An initial study by the NDP team has considered the feasibility of closer connectivity and integration. This is an area where technical solutions are apparent but considerable additional work is required to bring the security accreditation community on board.

7. MODULARITY AND OPEN SYSTEM ARCHITECTURE

The centre of gravity in Mission Modularity development to date has been in capabilities with no or limited warfighting roles, such as search and rescue, disaster relief, counter piracy. These missions require limited interface with a ships platform management or command system. Consideration of more complex capabilities such as mine countermeasures have tended to assume that the modular capability is essentially self-contained, again limiting the need to interface with IPMS or combat system. However, reaping the full benefit of Mission Modularity will require the rapid and seamless integration of module internal command systems with that of the host platform. This is of course already a feature of the Danish StanFlex system, where the module interfaces have been designed to connect to the standard Danish command system. However, in an environment where the fit of command systems across a fleet may not be standard, and moreso in a NATO environment where nationalities of modules and platforms mixes there is a need to develop and implement truly open architectures and – more significantly – the security protocols that allow this to be achieved. At a simple level it is possible that existing datalink systems (Link 16, 22 etc.)could be used in a hard wired form, but true interoperability will come when modules can be connected directly into the ship's command system. Initial work in the NDP indicates that this is complex but achievable. It is possible that a drive to a modular fleet could also be a significant driver to the fielding of truly open architecture command and control systems.

7. DEVELOPMENT OF STANDARDS

Mission Modularity will stand or fall on the delivery of standards for container design and interface specification. As such the delivery of effective standards and supporting design and operational guidance is critical to its implementation. In the UK this work began as the development within the NDP of the UK Container Specification. However, the need to ensure interoperability with NATO allies has seen this work merge with ST/MM's programme to deliver similar standards within NATO. Work to generate the initial release of NATO standardisation is now taking place within the NDP, ensuring that the full benefit of the UK's previous work is embedded within the NATO product and vice versa.

ST/MM is delivering two documents in the 2018-2020 timeframe. The primary document is a NATO standard issued under cover of a Standardization Agreement (STANAG) that will define the design and construction of containers for naval service, selection and location in containers of NATO compliant interface specifications, shock protection system requirements and additional requirements to enable safe and effective operation (for example requirements to provide information on weights and centres of gravity, requirements for air transport, etc.). An essential aspect of the standard is the specification and referencing were possible of existing commercial and military standards rather than the development of bespoke arrangements (which would drive cost). For example, the construction of containers following the DNV offshore service module standard.

Supporting the NATO standard is ANEP-91, an Allied Naval Engineering Publication (ANEP) that provides the supporting information required to enable mission modularity. The first edition of ANEP-91 was published in 2017 to provide initial guidance on mission modularity, with an updated edition scheduled for publication in support of the new standard.

ANEP-91 draws together the output from the various ST/MM, NIAG, specialist team and other studies that have been and are being conducted across NATO. For example, guidance on the design and integration of mission bays into ships, impacts of modularity on overall ship sizing, design and cost, logistics, personnel and maintenance aspects, and interoperability will all be included. The ANEP should be seen as the "how to" guide that supports the STANAG.

The UK Type 26 design team has worked closely with the NIAG study team developing the interface specifications to ensure that the developing material that will feed the STANAG is available for use. Liaison with the Type 26 design team has also been maintained with the designers of the shock mitigation system to ensure that the system is compatible with Type 26 and other surface ships, and vice versa. Interest is also being shown in adopting similar approaches in several US and European surface warship projects.

8. CONCLUSION

Mission Modularity offers the ability for navies to deploy flexible, adaptable ships capable of fulfilling a variety of missions whilst potentially delivering reduced fleet costs;

Effective delivery of a Mission Modularity capability requires a well understood set of interface specifications, especially when considered at a NATO level;

The degree to which Mission Modularity is applied needs to be balanced against the need to deliver inherent flexibility in a warship and the delivery of the vessel's core capability;

NATO has established a Specialist Team to address Mission Modularity, backed up by supporting teams in governments, navies and Industry;

The NATO Specialist Team on Mission Modularity will deliver detailed interface specifications and supporting guidance on developing and deploying a cost-effective modular capability effectively in a STANAG and ANEP in 2018.

9. REFERENCES

[1] *"Sea Change? Navies Make Progress in Delivering Modular Capability"*, Matthew Hood, RCNC, Janes Navy International, October 2014

[2] *"The Affordable Warship – a Design to Cost Approach based in the Concept Phase",* R F Lamerton, Thales Naval Ltd, UK, RINA Warship Conference 2007

[3] *"Versatility the key to Denmark's Evolving Navy".* Richard Scott, Jane's Navy International, October 1999.

[4] *Navy Littoral Combat Ship (LCS)/Frigate Program: Background and Issues for Congress*, Congressional Research Service Report January 2016

[5] *"Proposed UK Way Ahead on Modularity",* Dr M Courts, Dstl Naval Systems, January 2014

[6] *Standard for Certification No. 2.7-1 Offshore Containers*, Det Norske Veritas April 2006

10. AUTHOR'S BIOGRAPHY

David Manley holds the current position of Naval Authority Group Deputy Head, Ships Division in the UK Ministry of Defence. He is responsible for the development and management of the MOD's maritime combat safety programme as well as the Royal Navy's Survivability Strategy. His previous experience includes sixteen years working in maritime survivability and over twenty-five years working in the MOD. He is a DE&S Defence Engineering Senior Fellow, a Fellow of the Royal Institution of Naval Architects and a member of the Royal Corps of Naval Constructors.

11. ACKNOWLEDGEMENTS AND DISCLAIMER

The author would like to thank colleagues in DE&S, the SDA, Dstl, Navy Command Headquarters, NAVSEA, DMO in the Netherlands and the NATO Mission Modularity community for their help, support and encouragement in this work. The views expressed in this paper are those of the author and do not necessarily represent those of the Ministry of Defence

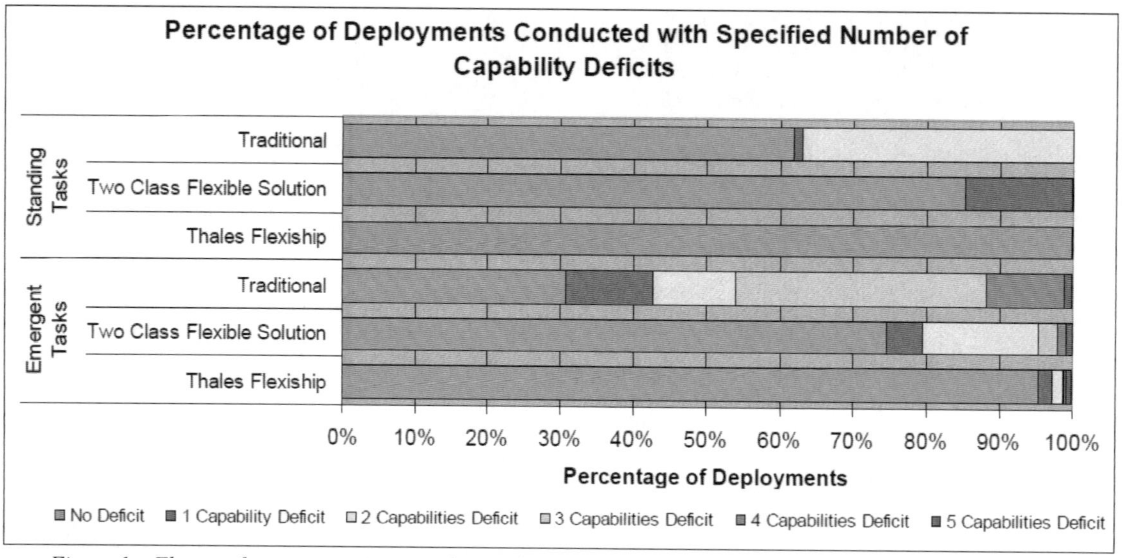

Figure 1 – Fleet performance against standing and emergent tasks: traditional vs. modular (Lamerton, 2007)

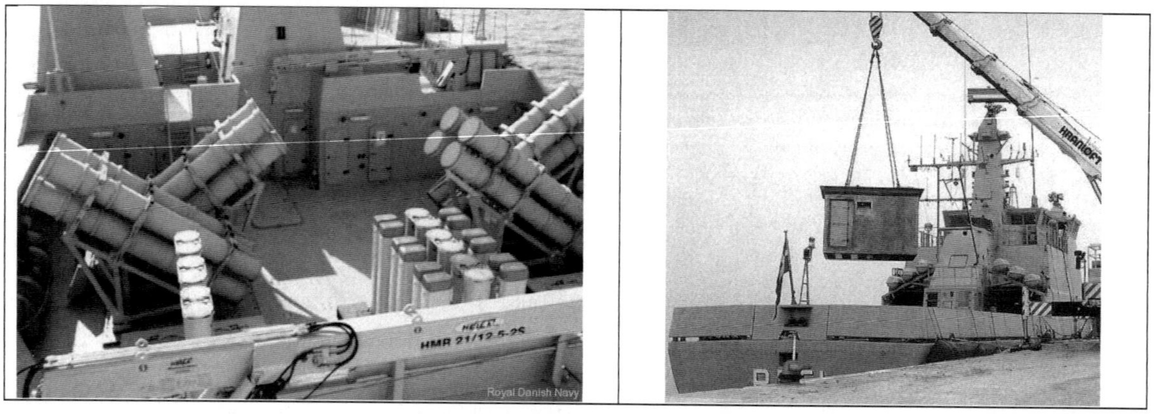

Figure 2 – The Danish StanFlex system; NATO's first example of mission modularity

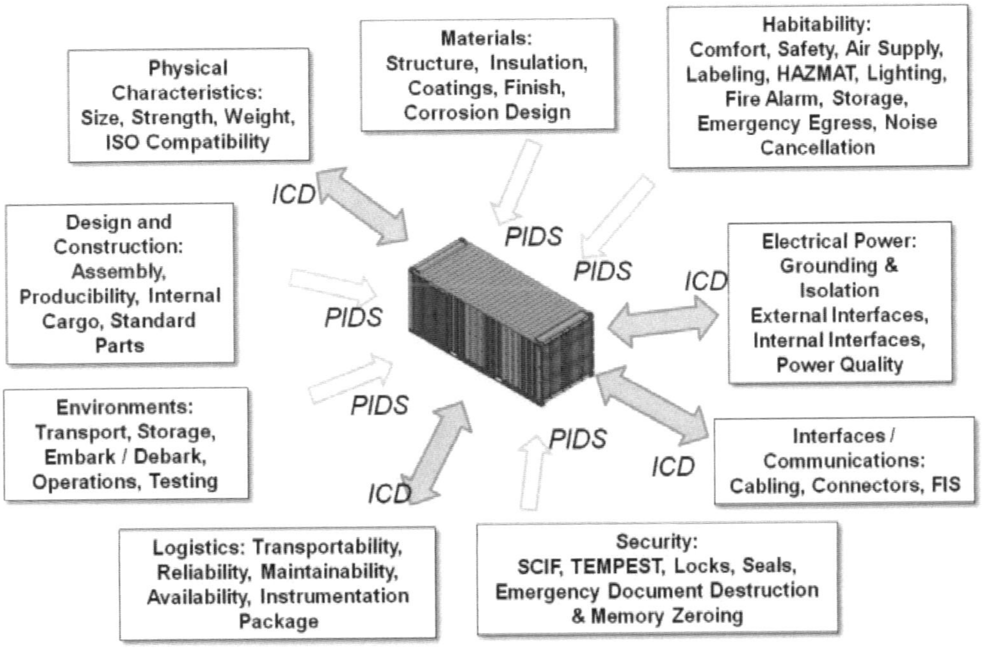

Figure 3 – The broad spectrum of modularity interfaces and issues

Figure 4 - LCS-2 Mission Bay

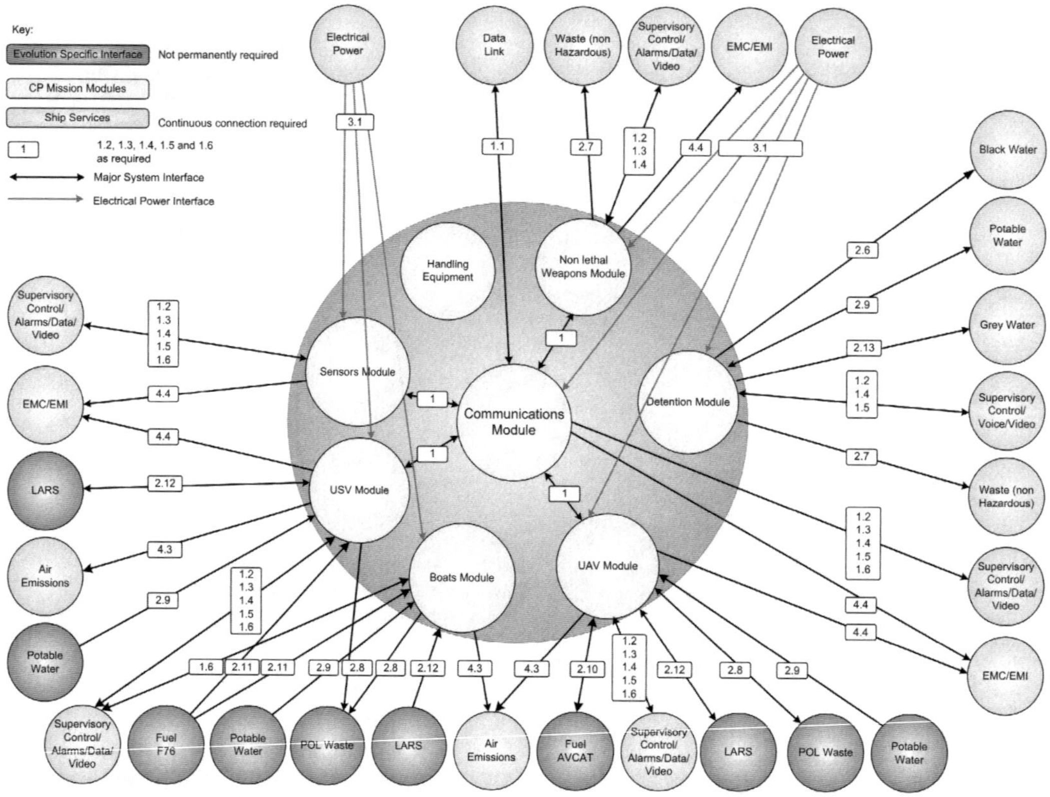

Figure 5 – Interface Diagram for a Counter Piracy Modular Capability (NIAG Study Team 168)

---

i *"Sea Change? Navies Make Progress in Delivering Modular Capability"*, Matthew Hood, RCNC, Janes Navy International, October 2014

ii *"The Affordable Warship – a Design to Cost Approach Based in the Concept Phase"*, R F Lamerton, Thales Naval Ltd, UK, RINA Warship Conference 2007

iii *"Proposed UK Way Ahead on Modularity"*, Dr M Courts, Dstl Naval Systems, January 2014

iv *"Versatility the key to Denmark's Evolving Navy"*. Richard Scott, Jane's Navy International, October 1999.

v *Navy Littoral Combat Ship (LCS)/Frigate Program: Background and Issues for Congress*, Congressional Research Service Report January 2016

vi *Standard for Certification No. 2.7-1 Offshore Containers*, Det Norske Veritas April 2006

THE NEXT GENERATION MCM PLATFORM – NOT YET FULL AUTONOMY

A J Aitken, BMT, Bath, UK

SUMMARY

Unmanned systems offer a revolutionary change to the concept of operations for naval minewarfare. Extensive research led BMT to conclude, that given the current rate of development of autonomous systems, a need shall remain for the foreseeable future for a specialist mine counter-measures (MCM) platform from which the autonomous systems can be hosted and operated. These ships shall form an integral element of a nation's maritime MCM capability and the new operating concept requires a very different platform to the existing ships. The paper summarises BMT's research into near-future autonomous minewarfare and examines the requirements of the next generation of MCM ships and how that is balanced against the need to provide an affordable, capable and safe national asset that can be adapted through life to continue to support the ever-developing autonomous MCM systems.

GLOSSARY

ASW	Anti-Submarine Warfare
ASuW	Anti-Surface Warfare
AUV	Autonomous Underwater Vehicle
CBRN	Chemical, Biological, Radiological & Nuclear
CFD	Computational Fluid Dynamics
CIC	Command Information Centre
CONOPS	Concept of Operations
LARS	Launch And Recovery System
LIDAR	Light Detection And Ranging
MCM	Mine Counter-Measures
MCMV	MCM Vessel
MIDS	Mine Identification & Disposal System
MLO	Mine-Like Object
nm	nautical mile
OA	Operational Analysis
OEM	Original Equipment Manufacturer
OPV	Offshore Patrol Vessel
RHIB	Rigid Hulled Inflatable Boat
ROV	Remote Operated Vehicle
UAV	Unmanned Aerial Vehicle
USV	Unmanned Surface Vehicle
UUV	Unmanned Underwater Vehicle (either remotely operated or autonomous
URN	Underwater radiated Noise
UxV	Unmanned vehicle (of any type)
VERTREP	Vertical Replenishment

1. INTRODUCTION

Sea mines continue to be a weapon of maritime warfare and an enduring threat to both military and commercial shipping but the means to detect and destroy the weapons is evolving rapidly. The desire to remove mine-warfare personnel from the mine threat has been long recognised and since the advent of remote vehicles, systems have been developed to provide that distance. Initially, offboard systems were used only for the mines' classification and destruction following detection by the mine-hunting ship's hull-mounted sonar. Fully autonomous systems now offer navies the opportunity to remove their personnel from the mine threat and to automate the dull and repetitive work of mine-hunting.

The design of mine-countermeasures (MCM) ships has not changed significantly over the last 100 years (Figure 1) but MCM technology is on the cusp of a revolution. Additionally, a large proportion of the word's legacy MCMV fleet is reaching the end of its operational life and requires replacement over the next decade. BMT began studies in 2015 to explore whether future mine-warfare would need a host platform and, if so, what the requirements for the platform would be.

Figure 1 – Examples of MCMV Designs Spanning the Last Century

This paper summarises the work done by BMT and presents the necessary characteristics of the next generation of MCM platform.

2. ESTABLISHING THE REQUIREMENT

2.1 THE FUTURE MCM TOOLBOX

Tethered unmanned underwater vehicles (UUVs) or remote operated vehicles (ROVs), as they are known, have been used by navies to identify and neutralise mines for several decades to distance the most risky MCM operations from the ship and personnel. Some navies are already using autonomous unmanned vehicles (UxVs) for domestic, coastal MCM operations and this trend is set to continue and their roles grow as the technologies develop. Exercise Unmanned Warrior conducted by the UK Royal Navy and managed by QinetiQ in 2016 demonstrated that a number of different autonomous technologies could be deployed and linked to conduct a joint operation.

The research team engaged with minewarfare experts, UxV manufacturers and combat systems suppliers to gain a thorough understanding of the latest MCM methods and technologies and the expected technological developments of MCM systems both in the short and long term. A synopsis of the different autonomous systems and their capabilities follows:

Larger, long range Autonomous Underwater Vehicles (AUVs) using a mix of side-scan and synthetic aperture sonars are being used to conduct wide area surveys of the minefield to detect mine-like objects (MLOs).

Smaller, shorter range AUVs are then deployed to identify, classify and, potentially, neutralise the objects identified in the initial search.

Unmanned Surface Vehicles (USVs) are anticipated to become the multi-role workhorses of the fleet fulfilling the roles of:

- Pulling a towed side scan sonar system. In shallower water, this is potentially a higher speed alternative to the large AUVs;
- Mine-hunting using an integral hull-mounted sonar;
- Taxiing smaller AUVs into the minefield/swept channel;
- ROV mothership used to deploy tethered inspection / mine disposal vehicles (potentially a manned role);
- Diver taxi – In a manual operating mode, the vehicle will carry clearance divers to an area of operation;
- Conducting acoustic and influence sweeps. Depending upon the threat, a sweep may be conducted in place of hunting. However, a sweep is expected to be conducted post the hunting operation to activate any undetected mines.

USVs for MCM are still developing but consensus amongst the OEMs is that they will be larger than the current generation to enable them to embark and power the envisaged payloads. A USV with a length of 13m, beam of 4m and displacement of 20 tonnes was selected as appropriate dimensions for a future-proof solution.

Unmanned Air Vehicles (UAVs) operating as part of the toolbox offers the ability to vastly increase its effectiveness. An UAV at height provides greater situational awareness. When equipped with LIDAR sensors, the UAV can make a rapid optical search ahead of the ship/USVs for floating/near-surface mines. It is hoped that the UAV will also act as a communications bridge between the ships and the USVs or AUVs on the surface, thus increasing their range and ensuring a communications link is maintained. Communications generally rely upon line of sight and there is a risk that a small vessel operating in large waves will lose visibility of the ship's antennae even at a short distance.

Data transfer from the AUVs to the ship is anticipated to happen either via the USV and UAV using a short range Wi-Fi link or directly between the AUV and ship via Wi-Fi/acoustic link (through-water) or on board the ship by direct connection. It is an aspiration that the AUV will surface and transmit directly to the UAV.

The paper "Generational Shift: How technology is shaping a step change in the future of mine counter-measures" [1] provides greater detail of the integration and control of the toolbox during MCM operations.

2.2 THE FUTURE MCM CONCEPT OF OPERATIONS

Legacy MCM practices require the ships to operate within the minefield but the desired Concept of Operations (CONOPS) envisages the MCM operation being controlled from a position outside the minefield and over the horizon; either from within the deployed Task Group or from shore (if host nation support is available). BMT identified four distinct CONOPS spanning the spectrum from the legacy approach to the fully autonomous concept (Figure 2):

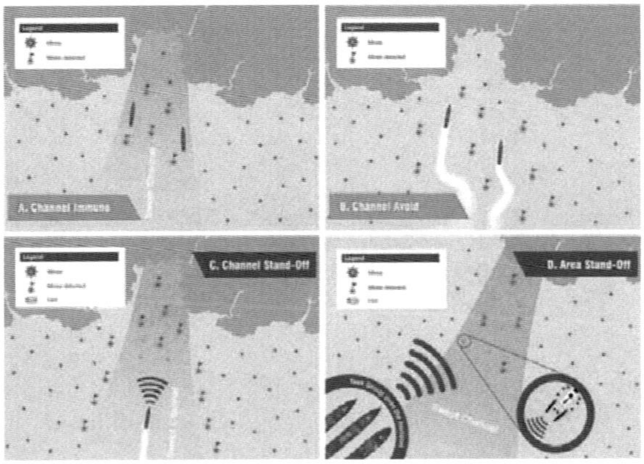

Figure 2 – The Four MCM Concepts of Operation

1. Channel Immune – A legacy practice that relies upon a highly capable ship which is effectively immune to the mine threat. The ships traditionally conducted towed minesweeping but most now have a hull-mounted mine-hunting sonar and tethered submersibles for mine-hunting.

2. Channel Avoid – The ship is still very capable but conducts mine-hunting operations only using a hull-mounted sonar and tethered submersibles. It is highly manoeuvrable and maintains a greater stand-off distance from the threat than the Channel Immune platform needs to.

3. Channel Stand-off – Autonomous systems are used to hunt and clear a swept channel but the limited maturity of autonomous systems technology requires the ship to follow the vehicles into the swept channel.

4. Area Stand-off – The MCM UxVs are deployed a considerable distance from the minefield and work autonomously to clear a channel through the minefield enabling shipping to pass through the channel upon completion. Tasking of the UxVs is conducted from a point over the horizon either on land or from a low capability ship.

Whilst the Area Stand-off CONOPS is the panacea most navies aspire to; several hurdles must be overcome before the adoption of long range, fully autonomous systems becomes a viable solution:

- Current AUV battery life limits their range and speed or operational time if there is a long transit to the minefield (particularly in bad weather);
- Legal limits on the use of autonomous systems and the need to keep a man in the loop. The IMO COLREGS [2] still require a maritime platform to maintain a visual lookout. Rules of engagement and the respect of international human rights laws mean armed forces still need a human with the most complete situational awareness available to make the decision to "pull the trigger" [3]. To achieve each of these requirements, effectively requires a live CCTV feed to be in place.
- It is desirable if the data being passed from the UxVs to the operators remains unprocessed. This has the advantages that the data is unclassified and the data and equipment on board the UxVs is of less value if the vehicle falls into opposition hands.

The disadvantage of the latter two requirements is that they vastly increase the quantity of data which needs to be passed back to the operator. Passing more data demands more battery life and also means that a high bandwidth, reliable and covert communications bridge is essential.

- Assurance must be provided that the area swept by the unmanned vehicles is clear of mines before a high-value military assets or commercial shipping transits the channel. A low signature, resilient MCM platform fitted with a hull mounted mine/obstacle avoidance sonar can conduct a low speed search of the channel as it follows the toolbox.
- Unmanned systems deployed far ahead of friendly units could be captured or suffer damage from hostile forces, and could become lost. The larger the UxVs (which scales with range), the higher value and less covert they become, thus increasing the protection, they themselves need.

Advances in mine technology means that the features of modern minefields are changing. Mine clearance operations are expected to become more onerous due to the need to hunt for more sophisticated mines, in deeper water and over large areas and clear long and wide safe channels. AUV and helicopter-based systems are unlikely to be able to provide sufficient time on station to give the necessary assurance to amphibious forces unless they are closely supported by a dedicated asset able to remain on task for extended periods (to maintain clear channels against mobile mines).

It is, therefore, Channel Stand-Off which appears to be the most likely CONOPS for the next couple of generations of UxV and hence the scenario upon which BMT focussed its research.

2.3 IDENTIFYING THE MCM TOOLBOX COMPOSITION

Discussions with unmanned systems suppliers and MCM equipment manufacturers established their products' capabilities and the probable make-up of a MCM toolbox comprising a mix of USVs, AUVs, UAVs and tethered UUV–based hunting/disposal systems. Using agent-based modelling as a simple Operational Analysis (OA) technique, the team simulated a range of mine clearance operations to identify the most appropriate toolbox composition to clear an example minefield within an acceptable timeframe. The different vehicles' ranges, speeds and survey windows were input and the following parameters were varied: number of each vehicle type; water depth; the size of the minefield being cleared; sensor range and overlap; neutralisation time; number of mines and mine-like objects.

The OA results showed that two survey units and two classification units provided an acceptable rate of clearance and detection and on a par with the clearance rates of legacy systems. The impact of additional systems was investigated but did not offer significant reduction in the operation's duration (Figures 3 & 4). Accordingly, it was decided the platform should embark the capability to deploy two medium AUVs simultaneously and two USVs with their range of payloads.

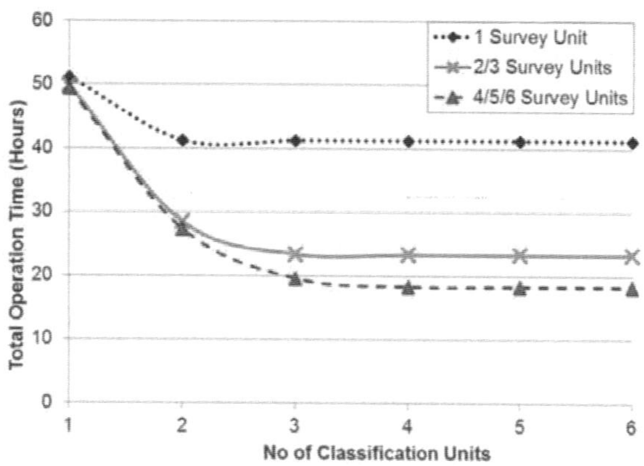

Figure 3 – OA Results Running A Range of Survey AUVs

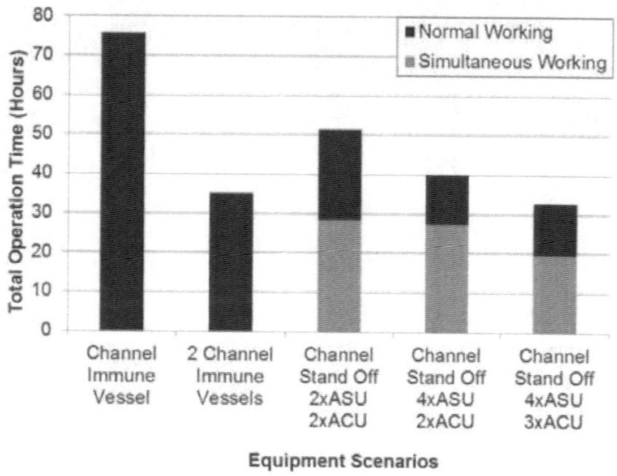

Figure 4 – OA Results Comparing Legacy Channel Immune Capability Clearance Time With Alternative Channel Stand Off Toolbox Permutations

2.4 DEFINING THE PLATFORM CAPABILITY

The agent based modelling also allowed the team to explore and identify where the MCMV needed to be in relation to the UxVs and the minefield. In turn this enabled the threat to the platform and its survivability requirement to be defined.

The model simulated realistic minefield geography and assumed that an 800m wide channel was cleared through a 50 nm deep field.[4] The toolbox was assumed to clear a 5-10 nm section ahead of the ship. This range matches UxV suppliers' and MCM experts' expectations of near-future MCM systems' capabilities. The distance provides a reasonable balance of transit time versus survey time for the vehicles' batteries. The ship is expected to remain within the channel during clearance operations and follow the toolbox at a safe distance.

The ship's anticipated location during MCM operations means that its acoustic, magnetic and electrical signatures should be no worse than a front line surface combatant.

The ship may operate within the confines of the swept channel for several days so must be highly manoeuvrable and capable of holding its position and heading to a high degree of accuracy.

The MCM platform will be the first platform to transit the swept field. Therefore, if there is an undetected mine, it is likely to be the asset which triggers it. With the risk that further undetected mines may be present, the platform needs to have sufficient damaged stability to survive such an incident and the shock capacity for its systems to function sufficiently for it to self-extract itself from the channel and not require another vessel to be placed in danger to effect a tow.

The OA made clear that this ship would need to have the capabilities of a warship; accordingly it should be designed and constructed to a classification society's naval rules set. The ship is designed to satisfy ANEP-77 [5] with munitions safety designed to Def Stan 00-101 [6] standards and underwater signatures satisfying AMP-14 [7]. This rationale aligns with the duty of care Ministries of Defence have to protect their personnel and provide them with equipment appropriate to the threat they are exposed to. Satisfying the Combat Safety Case is particularly pertinent to the UK [8] but other nations are adopting the approach.

The OA showed that while conducting MCM, the platform would often be operating alone and up-threat and probably outside the task group's protective umbrella so will need to have sufficient self-protection against the anticipated threats of low, slow flyers and fast inshore attack craft (FIAC), both manned and drone, in particular. In the littoral environment, the ship may be within range of land-based artillery.

MCM is a key element of an amphibious operation so the ship is expected to deploy as part of a task group and must be able to maintain passage speed. Additionally, navies may find a platform of this type useful for patrol duties so a cruising speed of 15 knots and a top speed of at least 17 knots were deemed the minimum necessary. A minimum range of 3500nm and an endurance of 30 days were specified. These values are typical of a warship this size and enable the platform to sail as part of the task group or deploy unsupported to transit to expeditionary operating locations.

A total complement of at least 60 persons was set initially to enable ship sizing. Subsequent studies were conducted (and continue) to establish both the ship's complement and the MCM component. The MCM crew needs to be sufficient to plan operations, support and sustain the toolbox, and compile and maintain the MCM picture.

2.5 THE NEED FOR A DEDICATED MCM PLATFORM

The Channel Stand-off CONOPS is envisaged to be the maximum realistic level of capability possible for the next few generations of UxV so a mothership for the unmanned systems will be necessary to provide a persistent, offensive MCM capability.

Many are questioning why this needs to be a dedicated MCM platform when MCM systems are becoming more modular? A dedicated platform does not detract from the function of the rest of the fleet.

If the MCM capability is embarked on a frigate or destroyer, platforms which have the necessary survivability, then the ASW/ASuW protection of the task group would be diminished and a high value asset is placed in harm's way. Also, many combatant platforms lack the manoeuvrability and the space to accommodate the systems.

Offshore Patrol Vessels (OPVs) lack the survivability features and hull shape to be a safe and suitable platform for MCM operations and handling large UxVs.

Naval auxiliaries are often too large to be covert and tasking them on MCM would detract from the support role they provide to the wider task group. Naval auxiliaries (and commercial platforms of opportunity, should they be considered) are usually built to commercial standards so lack the inherent capability and the signature controls to be safe for the Channel Standoff CONOPS.

3. THE SOLUTION - THE FUTURE MCM PLATFORM

3.1 PRINCIPAL PLATFORM PARTICULARS

To accommodate the MCM toolbox, the 60 person complement, the ship's systems, stores and tanks, a ship with a waterline length of 85m and a beam of 16.8m is required. The ship's maximum draught was limited to 5.5m to enable it to operate in shallow waters around the entrances of smaller ports and harbours. The ship is arranged with its accommodation and operational spaces forward, leaving the aft half of the ship as a semi-enclosed working area for stowage and support of the MCM toolbox.

The ship will be constructed of mild steel built to naval ship standards, giving an all up displacement of ~3,500 tonnes. Steel was selected as it enables the ship to be built in any indigenous shipyard and does not necessitate the specialist skills and facilities required for composite construction. A steel hull is also significantly cheaper than a similar sized composite platform. It may be possible to construct the superstructure of composite to reduce top-weight, if necessary.

The beam is driven by the breadth of the USVs and the need to be able to service and support them. Achieving naval damage stability requirements also influences the beam.

3.2 HULLFORM

The hull's form was developed to provide a solution which best delivers the numerous requirements placed upon the platform:

- It must provide a stable and low motion platform during MCM operations when station-keeping behind the toolbox and launching and recovering the large USVs. This requirement favours a deeper, heavier, fuller hull.
- The military speed requirement drives toward a finer hull with LCB aft of amidships. However, accommodating the open working deck and AUV garage aft drives the LCG forward.
- Many navies may consider using the platform for a hydrographic role so producing a hullform that generates low self-noise at survey speeds was a primary requirement. Consideration of the bow shape, thruster positions and identifying locations for sensors were key to achieving this.
- The ship must meet naval stability standards.

The ship's open aft end is also a challenge to obtaining sufficient buoyancy to meet the naval damaged stability requirement and is a driver of the main deck height.

The hull was derived from a previous similar design which had been model tested and analysed for seakeeping and manoeuvring performance. The baseline hull was compared with BMT's database of ships with similar parameters to identify improvements to the hull to minimise resistance. A number of options were analysed using FineMarine CFD and changes showing most promise were incorporated into the final hullform. The CFD analysis was also used to examine the streamlines in way of the sensor positions to reduce risk of interference.

3.3 MACHINERY ARCHITECTURE & MECHANICAL SYSTEMS

The ship is provided with a diesel-electric 690V power generation and propulsion architecture which is split equally between two main engine rooms. Each machinery space contains two acoustically mounted and enclosed diesel generators which are connected via the switchboards to acoustically quiet motor/gearboxes. The multiple prime mover solution offers the flexibility required for the variable power demand and the redundancy necessary to meet the survivability requirement.

Depending upon the installed power required, the signature requirement and, potentially, customer preference, the generators may be either high speed or medium speed.

The ancillary and auxiliary pipe systems and their pumps/motors will be sized, specified and arranged to minimise the creation of noise.

The ship shall comply with MARPOL requirements [9]. For emissions and pollution control, this is achieved via the size and distribution of the fuel tanks. A water-ballast treatment plant will be provided to prevent the transfer of invasive waterborne species. For a shock rated vessel, a challenge is identifying an exhaust gas scrubber which is capable of withstanding shock. Most scrubbers use ceramic filters which risk shattering and falling into the engines when shocked. Military spec scrubbers with a bypass or alternative solution must be identified.

3.4 POSITION KEEPING & MOTION CONTROL

A conventional twin shaft and rudder arrangement with lateral tunnel thrusters was selected as the only viable solution for a MCM platform as the system must be shock certified and have low acoustic and electric signatures. Azimuthing thrusters are not typically shock capable and azipods place the electric motors/gearboxes outside the hull which compromise their signatures. The system will probably include controllable pitch propellers to enable rapid direction changes but their shock capacity and noise levels compared to a fixed pitch propeller will be evaluated prior to selection.

A dynamic positioning system equivalent to the DP2 standard would offer the system redundancy necessary for MCM operations of this nature. The ship is likely to be operating in areas with strong tidal currents and must be able to hold against cross winds/wave so is fitted with large tunnel thrusters fore and aft.

A flume-type roll reduction tank is provided beneath the Bridge to reduce the ship's motions whilst stationary and at low speed. Providing a stable, comfortable platform is necessary not just to make launch and recovery of the offboard assets (UxVs, boats and other equipment) easier but also to create an effective environment for the ship's staff to analyse the gathered data, compile the situational picture and plan the mission.

The ballast system will include anti-heel and trim-control tanks to maintain the ship upright and level during launch/recovery of the MCM payload, particularly the laden USVs.

3.5 SURVIVABILITY FEATURES

Mines can be triggered by a number of means; these are typically changes in the water pressure, electrical or anomalies in the earth's magnetic field, by recognising an acoustic profile or by direct contact (e.g. the conventional horned mine). An acoustic signature may activate a mine but, due to sound's ability to propagate through water, it is difficult for the mine to establish the target's location and direction of travel unless there are multiple microphones/hydrophones (which is becoming more common in complex minefields). Magnetic, electrical and pressure signatures provide an easier means to locate the target. Mines normally use more than one type of detection sensor. Typically they are primed by URN signal, and then actuated by the magnetic signature.

A well configured multi-coil 3D degaussing system can reduce a ship's magnetic signature by up to 90% [10] and is considered essential for a MCM platform which will be operating within the cleared channel for extended periods (and depending upon wind, current and interaction with the USVs may be operating up to the channel's edges).

As discussed above, main machinery will be acoustically mounted and ancillaries specified and mounted to minimise noise creation. Systems and connections will be designed and installed to minimise noise creation and avoid noise shorts (e.g. by using flexible couplings and connections).

The ship is expected to operate at a very low speed or be station-keeping when it is in the swept channel which will mean pressure changes caused by the hull will remain low.

Despite the underwater signature reduction measures provided, there remains a possibility that a MCMV will encounter an underwater shock load in its lifetime. The likelihood and load experienced by a Channel Stand-Off MCM should be lower than the legacy systems. The ship structure will be designed to withstand the load specified and the ship's propulsion system is designed with sufficient shock capacity to enable it to self-extract from the threat area following a shock event. Other equipment and systems will be designed to remain captive at least up to the hull lethal shock load. It is anticipated that the ship will have a shock capability equivalent to a frontline surface combatant warship.

The ship's above water signatures will be minimised by a number of means. The ship has a low profile which will reduce its visual signature; its superstructure and mast shall be inclined and dihedral corners avoided wherever possible to reduce its radar cross-section; 'cheese-graters' will be fitted to the main engines' exhausts to reduce the infra-red signature.

In common with most military platforms, the ship is designed with a citadel and HVAC system to offer protection to the personnel in the event it needs to transit a CBRN environment.

Fire protection, fire-fighting and damage control systems and arrangements are designed in accordance with ANEP-77's requirements to provide naval standard protection with which crews shall be familiar and offers protection against the perceived threat to the ship whether it is operating alone up-threat or within the task group. The following assumptions underpin the philosophy adopted for the design of the ship's damage control and protection systems:

- The ship will be operated by a core sailing crew which will be supplemented by different sized mission crews as required for the ship's particular deployment, whether MCM or an alternative role.
- All personnel embarking the ship are assumed to be damage control trained and shall support a damage response as required.
- The level of response to a particular incident shall be a command decision and depend upon the number of personnel available to support the effort.

3.6 EMBARKING THE MCM TOOLBOX

Unmanned technologies are evolving fast so the ship (which is designed for a life of at least 25 years) is expected to embark several generations of the MCM toolbox over its lifetime. Hence, it is fundamental that the ship's design is able to readily accommodate these different solutions without incurring significant cost or impact to its availability.

To achieve this flexibility, an approach was adopted similar to that used by the offshore sector where the ships are adapted frequently to embark and deploy different project payloads. This approach was helped by most navies' assumption that MCM is moving towards a modular solution that enables the capability to be transported by air and road and deployed from shore as well as the MCMVs. The solution is ensuring that sufficient space with ready over-side access and the means to embark, support, deploy and recover the different toolbox components is provided.

A large open working deck would provide the best solution for the surface and underwater vehicles in terms of space, flexibility, crane access and water access. However, enclosing the deck to form a garage is highly desirable for operational reasons: it provides a protected, illuminated and discrete environment to maintain, support and re-role the vehicles. The space is served with overhead cranes to enable movement of the AUVs, USV payloads and vehicle spares. The space is adjacent to the MIDS magazine to facilitate easy movement of the armed MIDS vehicles to the USVs or point of launch (in the event the system can be deployed from the ship). Several other features which must be adjacent to the garage include: workshops, battery charging facilities, spares and consumables storage, and storage for the USVs' payload modules. The MCM capability will continue to include clearance divers for the foreseeable future to support the identification and recovery or disposal of unfamiliar munitions or munitions which the UUVs are unable to reach. Therefore, the mission space needs to include or embark the divers' facilities. A crane is provided to support deployment/recovery of the sweep system components towed by the USVs and can also be used to embark the mission containers when alongside.

The larger, long range AUVs may be deployed directly from the ship and it is intended that a dedicated, gantry-type launch and recovery system (LARS) is provided adjacent to their stowed positions. Alternatively, the AUVs may be deployed and recovered via the ship's crane.

Determining whether the solution for the USV LARS should be davits or a stern ramp was the subject of much debate. A davit solution was selected for the following reasons:

- A stern ramp removes significant buoyancy from a critical position in the ship;
- A stern ramp requires the ship to be on a specific heading relative to the wind/waves/current and is easier if the ship is making way. This may be hard to achieve within the confines of the swept channel [11].
- Keeping the USVs at the stern of the ship reduces crane access for re-roling and servicing the craft. Also, a hoist system would be required to bring the USVs level to make re-roling and servicing easier;
- A single stern ramp is a single point of failure preventing deployment/recovery of the USVs and potentially, jeopardising the mission. Accommodating two adjacent ramps in the MCMV's transom is difficult – an off-centre ramp will have a reduced operating window in large waves; integrating the ship's steering gear is difficult; more buoyancy is lost from the ship's stern;
- If the ramp is shared by numerous USVs then the USVs must be moved forward of the ramp to enable others to use it, having the effect that greater ship length is required for the same capability as davits;
- The need to immerse a ship's transom to accommodate a stern ramp results in a much less efficient hull form resulting in higher fuel usage and associated operating cost.

A davit based solution is not without its challenges. The USVs are large and heavy and sufficient structure must be provided to support them. In order that there is redundancy in the davit solution, a boat handling system is needed to move the USVs between the davits.

A challenge to both systems is recovery of the unmanned vehicles. It is undesirable, from a safety perspective to put a man in the loop (i.e. on board the USV) to effect its recovery. Capture systems for USVs are becoming more common but connection of the davit hooks to the USV lifting points needs development and is an area both davit and USV OEMs are investing effort. The USV's aft davit point may interfere with its aft working deck and the payload modules must accommodate this. A major challenge facing stern ramp solutions is hitting the quiescent period of both vessels' relative motions to recover the USV at the right attitude and without causing damage.

The UAVs are provided with a hangar with direct access to the flight deck. Up to 3 UAVs may need to be embarked to support 24 hour working depending upon how they are used and what alternative solutions for communications between the ship and USVs are achieved. The flight deck may also serve as a VERTREP position for manned aircraft.

3.7 OPERATIONAL SPACES

The MCMV will be "fought from the Bridge". As is increasingly common on lean manned, minor warships, the Bridge is the centre for situational awareness. Greater automation of warships' engineering and combat systems and their links to ship's IPMS and CMS allows fewer people to be concentrated on the Bridge and still provided with the full situational picture.

Central to ensuring command effectiveness for the MCM mission is the link between the Bridge and the MCM mission planning and direction. Therefore, a Command Information Centre (CIC) is provided immediately aft of the Bridge. The CIC will contain a number of consoles for MCM mission planning, control of the unmanned vehicles, operation of the ship's guns, control of the ship's machinery and systems, external communications.

Near to the CIC will be a separate, quieter space where the data returned from the UxVs shall be analysed and the MCM picture is compiled.

The intention is that the mission consoles/systems in the CIC and analysis/compilation spaces and the various associated servers shall be modular enabling them to be reconfigured or replaced if the ship is required to fulfil alternative, non-MCM roles. The modular approach also makes them readily upgradeable as technology advances.

3.8 ACCOMMODATON SPACES

A flexible accommodation solution shall be necessary to accommodate the variable crew numbers and compositions which will be embarked for different missions. Some MCM missions shall be more personnel-heavy than others and non-MCM duties will require a different composition again. Also, the unknown ratio of male/female staff must be accommodated efficiently.

The accommodation comprises a mix of single, twin and four berth cabins to a high, modern, naval standard. The standard can be relaxed if higher occupancy cabins are desired by a particular client. As is necessary on a MCMV, all the accommodation is located above the waterline.

To meet the flexibility need, the berthing accommodation is designed for regions to be shut down if they are not required. A single galley and dining hall is the easiest to support with a small catering staff and allows a small crew to come together socially. Separate recreation rooms are provided for each rank/rate as experience has shown that crew and officers do prefer some separation even on a lightly manned ship [12].

3.9 COMBAT SYSTEMS & COMMUNICATIONS

The weapons, sensors and combat management system fitted to the platform will depend upon the customer's requirement based upon the threat he foresees the ship being exposed to and how it shall be operated. The design can therefore accept a range of defence systems from small local, point-defence weapons up to a 57mm gun integrated with an all-weather fire control director.

The ship is expected to need a surveillance radar capable of maintaining an operational picture of the unmanned assets and this would also allow detection of incoming threats.

The communications fit would also be subject to customer preference and space is made available on the bridge top and mast for a variety of antennae. Space within the ship and margins in the systems are provided to accommodate the necessary equipment.

Key to the platform's success as a next generation MCMV is a system which links the unmanned systems and the ship's combat management system. For this reason, the platform is assumed to be equipped with an intelligent infrastructure technology which would allow a small number of personnel to plan and execute a mission in a dynamic operational environment, employing a large number of unmanned vehicles and deployed sensors from any equipment supplier. Such a system would enable co-ordination of concurrent operations spread over a wide area. A system of this type would allow control of different vehicle types from different manufacturers through a single, common interface and compilation and presentation of the MCM situation in concert with the wider operational, combat picture. The system should be able to integrate fully with the rest of the task group to assist creation of a comprehensive situational picture.

An equipment agnostic, open architecture system simplifies integration activities when new applications are added to the mission system and reduces the maintenance burden and logistics chain by reducing the variety of mission system hardware components on board the platform. A system incorporating international military standards for all information exchanges would ease international interoperability too.

4. ALTERNATIVE NON-MCM ROLES

A small, capable, low signature and highly manoeuvrable platform with good situational awareness facilities and adaptable mission spaces and systems is likely to be highly attractive to navies for a variety of roles.

The platform's inherent survivability features allows it to be deployed safely into a higher threat environment than an OPV can be.

4.1 HYDROGRAPHIC SURVEY

Many of the ship's inherent features and the MCM tools embarked enable it to support up-threat rapid environmental assessments (REA) ahead of other vessels' passage. For this reason the platform can be fitted with a multi-beam echo-sounder in addition to the mine/obstacle avoidance sonar.

The open aft working deck lends itself to fitting an A-frame on the transom and stowing towed systems.

The USVs may be replaced or reconfigured to fulfil the role of manned or autonomous survey motor boats.

4.2 MARITIME SECURITY & INTERNATIONAL ENGAGEMENT

The ship suits the requirements of a patrol and constabulary role. The UAVs and USVs can enhance the platform's situational awareness. The ship is provided with two 7.5m RHIBs to support low threat interdiction operations. The USVs may be replaced with specialist interdiction craft to enhance the capability.

The mission space may be used to embark Humanitarian Assistance and Disaster Relief (HADR) modules or provide shelter to rescued/evacuated civilians.

4.3 ANTI-SUBMARINE WARFARE (ASW) SUPPORT

The ship's controlled signatures and space to embark a variable depth towed array sonar suite enables it to deploy as an effective ASW surveillance capability. The platform would not have the full capability of an ASW frigate but can bolster monitoring and deterrence in a nation's home waters allowing the high-value specialist frigates to deploy on more demanding operations. Deployment of thin-line towed array sensors from the USVs could augment the surveillance picture.

5. CONCLUSIONS

The adoption of autonomous systems for mine countermeasures is causing a revolution in the approach to mine warfare. While autonomous systems offer the opportunity to remove the man from the minefield during MCM operations; the systems are not yet capable of delivering a persistent, offensive MCM capability.

BMT's research showed that for the next few generations of UxV, a dedicated, capable MCMV shall be required to act as a mothership to the unmanned systems. This new platform shall be very different to the legacy MCM fleet to allow it to embark and safely deploy the UxV-based toolkit of underwater, surface and air vehicles.

The expected Channel Stand-Off CONOPS will require a highly manoeuvrable, signature-controlled platform to follow the deployed UxVs of the MCM toolbox into the swept channel. OA conducted by BMT showed that the platform and its assets shall have the capability of two legacy MCMVs. However, the lower mine threat which the platform will be exposed to will enable it to be constructed of steel. A steel hull is significantly cheaper than a composite platform of similar size and does not require the rare and specialised facilities of composite construction. Hence, the platform may be built indigenously by prospective nations.

To remain useful over its lifetime, the platform must be readily adaptable to accept each new generation of the MCM toolbox, whether this is a change of the vehicles or the systems and sensors which will control them. The platform's size, capability and adaptability will make it a versatile asset in a navy's fleet able to fulfil numerous alternative military roles.

Several challenges still exist to delivering the Channel Stand-Off CONOPS successfully, including data transfer between the UxVs and the ship and recovery of the USVs, but the vehicle and equipment suppliers are working to deliver solutions. These are not envisaged to be major challengers to the need for a platform or to its design.

6. ACKNOWLEDGEMENTS

The author would like to thank his design team colleagues at BMT. The valuable input from the numerous UxV and marine systems suppliers, including Atlas Electronik, ECA Group, Northrop Grumman, Thales and Wartsila. Special thanks to QinetiQ for its support and collaboration on many elements of the MCM research.

7. DISCLAIMER

The opinions expressed in this paper are those of the author and do not necessarily represent BMT.

8. REFERENCES

1. Generational Shift: How technology is shaping a step change in the future of mine counter-measures. J C Rigby, J McWilliams, J Johnson, INEC 2018

2. Convention on the International Regulations for Preventing Collisions at Sea, 1972 (COLREGs)

plus amendments, International Maritime Organization, London.

3. Source: International Committee of the Red Cross, https://www.icrc.org/eng/resources/documents/interview/2013/05-10-drone-weapons-ihl.htm

4. Naval Mine Warfare, NWP 3-15, US Navy, March 2004

5. The Naval Ship Code, Allied Naval Engineering Publication, ANEP – 77, Edition G Version 1, NATO, September 2017

6. Def Stan 00-101, Design Standards for Explosives Safety in MOD Ships and Submarines, Part 1 - Surface Ships, UK Ministry of Defence

7. AMP-14, Underwater Signature Range Information for NATO Mine Countermeasures Vessels, North Atlantic Treaty Organization, Brussels

8. Combat Safety and Survivability: Using Survivability Techniques to Assess Crew Safety During Combat, D Manley, RINA Warship 2016, Bath

9. International Convention for the Prevention of Pollution from Ships (MARPOL), 1973 plus amendments, International Maritime Organization, London

10. Efficient Re-degaussing Technique for a Naval Ship Undergoing a Breakdown in Degaussing Coils, Dong-Wook Kim et al, Journal of Magnetics, The Korean Magnetics Society, May 2016

11. Boat Launch and Recovery – A Key Enabling Technology For Flexible Warships. Kimber A. Pacific 2012, Sydney: IMC

12. Anecdotal evidence provided by crew of BNS Castor at DSEI 2015.

9. AUTHOR'S BIOGRAPHY

Alex Aitken is a Principal Naval Architect at BMT and runs the Surface Ships Outfit & Arrangement Team. Alex has led the platform aspects of BMT's MCM research since 2015 and is Design Manager of its VENARI-85 MCM platform.

THE IMPLICATIONS OF ADAPTABILITY, FLEXIBILITY AND MODULARITY FOR ROYAL NAVY WARSHIP DESIGN

M. Courts, BAE Systems Maritime – Naval Ships, UK
C. Broadbent DSTL Platform Systems Above Water Group

SUMMARY

The paper summarises the results from a series of studies performed by BAE and DSTL on the implications of Adaptability, Flexibility and Modularity on RN warship design in terms of their main advantages and disadvantages. It considers the design strategies that contribute to achieving these characteristics. It reviews these approaches and the interdependencies between them and identifies the benefits that can be expected in terms of production, operation and through life upgrade. Reference is made to studies performed by BAE and DSTL which attempted to quantify the potential effect of some of the design features on representative warship concepts. Payload modularity is shown to be a subset of, and to an extent dependent upon, many of the identified features. The possible benefits of payload modularity are discussed and a potential strategy for future RN warships identified. The paper concludes with the results of a study on a range of combatant concepts with varying degrees of flexibility and multi mission capability, using the BAEFASIP™ tool to identify the optimum fleet mixes. This showed that the right mission capability combinations in potentially larger designs can lead to a more cost effective overall fleet mix than traditional more minimal ship concepts.

NOMENCLATURE

AAW	Anti Air Warfare
ASW	Anti Submarine Warfare
AUV	Autonomous Underwater Vehicle
BAE	BAE Systems Maritime – Naval Ships
C2	Command and Control
C4I	Command, Control, Communications, Computers and Intelligence
CIWS	Close In Weapon System
COTS	Commercial Off The Shelf
CTD	Conductivity, Temperature and Depth
DSTL	Defence Science & Technology Laboratory
FASIP	Functional Analysis of System Implementation Parameters
FBNW	For But Not With
FOC	First Of Class
HADRO	Humanitarian Aid and Disaster Relief Operations
ISO	International Standards Organisation
L & R	Launch & Recovery
MCM	Mine Counter Measures
MOTS	Military Off The Shelf
OEM	Original Equipment Manufacturer
SCG	Small Calibre Gun
SSM	Surface to Surface Missile
UAV	Unmanned Air Vehicle
USV	Unmanned Surface Vehicle
UXV	Unmanned Vehicle (generic)
VLS	Vertical Launch System

1. INTRODUCTION

1.1 BACKGROUND

Affordability of Naval vessels is a growing issue as government budgets come under more pressure. As a result alternative design, procurement and operation strategies are being investigated, and adopted, around the world to try and close the gap between aspiration and affordability. A frequent demand from politicians is to perform more with less and the concepts of Adaptability, Flexibility and Modularity are seen as the keys to achieving this, although generally with no clear definition of how. As a result a series of studies have been performed at BAE Systems Maritime Naval Ships (BAE) and the MOD's Defence Science and Technology Laboratory (DSTL) to understand the implications of Adaptability, Flexibility and Modularity on the design and cost of Royal Naval Ships and the sort of options they may offer for the future shape of the Royal Navy (1).

1.2 STUDIES

The studies have covered several elements

- **Definitions** – Developed to clarify the terminology commonly used.
- **Decomposition** – Breakdown of the complete warship philosophy from system property objectives through design strategies to individual features that affect the ship configuration, operation and cost. The inter-relationships and dependencies between all these elements were established. Design studies were used to quantify the effects of applying specific design principles.
- **Combat System Modularity** – The implications of combat system (payload) modularity in particular were examined and a potential future strategy for the RN proposed.
- **Alternative future concepts** – Concept designs to meet a range of capability requirements exploiting payload modularity were generated and costed.
- **Fleet structure optimisation** – A fleet mix cost effectiveness optimisation study was performed and used to identify the most promising ship concepts.

2. DEFINITIONS

Much has been written on Modularity, Flexibility and Adaptability in the context of warship design, build and operation. Unfortunately ambiguity has arisen in some quarters, as terms have been applied in the context of both ship build and operation, with a different meaning in each case. There is thus a need for a clear set of definitions. Accordingly three categories of terminology have been established as follows:

- **Capabilities -** The terms used to specify military objectives, defined in Table 1.
- **Entities -** Physical objects referring to parts of a ship and the ways in which it is built, defined in Table 2.
- **Characteristics -** The properties of an entity that it is desired to maximise, defined in Table 3.

Table 1: Capability Definitions

Capabilities		
Term	**Description**	**Examples**
Role	The fulfilment of a military objective.	- Area Anti Air Warfare. - Anti piracy.
Operation	A series of actions undertaken in a specific scenario to meet defined objectives	- Escort shipping under threat from a hostile air force. - Patrol an area to deter piracy.
Function	The purpose and actions of a physical part of the ship in supporting the ship's operational role. Functions can be arranged in a hierarchy with lower level sub functions supporting top level primary functions.	- At the highest level the primary warship functions are generally identified as ▪ Fight ▪ Move ▪ Float ▪ Hotel - At lower levels the functions can be broken down into sub functions for example ▪ Air surveillance ▪ Engage surface target ▪ Provide casualty treatment ▪ Etc.

Table 2: Entity Definitions

Entities		
Term	**Description**	**Examples**
System	A collection of components and assemblies, which together form a complete whole implementing particular functions. Just as functions can be structured hierarchically so can systems. There is generally a direct mapping between systems and sub systems and their associated functions and sub functions.	- Anti air warfare system comprising ▪ Sensors. ▪ Command & control. ▪ Effectors. - Fuel oil distribution. - Fire-fighting.
Interface	The common boundary between systems or sub systems or between functions or sub functions. It refers to 'hard' physical properties', soft' data or information and human factors.	- Fuel oil supply to an engine - Air conditioning demand of an operational space. - Data requirement of a command system.
Ship	A complete self-contained operating warship.	- Frigate.
Spatial unit	A measure of the size of subdivisions of the total space available in the ship, which may be different in different directions.	- Frame spacing. - Compartment length. - Longitudinal bulkhead separation.
Zone	A region of the ship ideally made up of an integer number of spatial units. The elements contained within the region share a common operational or functional characteristic.	- Damage control zone. - Accommodation zone. - Command and control complex. - Adjacent propulsion machinery spaces.
Component	The smallest element individually identified on the bill of materials, or manufactured and handled, within the shipyard.	- Steel piece part. - Pump. - Engine.

Entities		
Term	**Description**	**Examples**
Assembly	A self-supporting collection of outfit components and or enclosing/supporting structure that is manufactured, assembled and completed away from the ship in a workshop environment. It may itself comprise sub-assemblies. There is not necessarily a correlation between an assembly and a system or function although it is possible for a system or function to be self-contained within an assembly.	- Mast. - Bulkhead. - Deck panel.
Module	An assembly that is so sufficiently complete that it can perform an identified discrete function or sub function in support of the ships operational capability. Generally assumed to be easily removable from the completed ship and capable of operation away from the ship provided appropriate services are provided via unchanged interfaces on the module itself.	- Plug in board in a computer system. - Air conditioning plant on a bed plate. - Weapon and supporting equipment in a box.
Unit	Structurally complete sub-assemblies of a block that contain one or more compartments.	- Mast. - Section of double bottom.
Block	The largest structurally complete production sub assembly prior to complete ship assembly at berth.	- Bow section of destroyer. - Aircraft carrier blocks.
Variant	A design that has a large degree of commonality with a baseline design.	- A ship lengthened by the insertion of an additional segment into its centre with the original ends retained. - A ship identical to an original but with an alternative machinery system.

Table 3: Desirable Characteristic Definitions

Characteristics		
Term	**Description**	**Examples**
Producibility	A measure of the ease with which a design can be manufactured assembled and tested. This may be enabled by either or both reducing work content and improving the efficiency of the work required. This may be achieved by breaking items down into sub-assemblies that can be built and tested separately before coming together or by re-designing assemblies to have fewer components requiring little effort to assemble.	- Heavier structure with few components and weld lengths or flatter panels with less curvature.
Operability	A measure of the ease with which a ship or system can be operated by its crew.	- Ship that has easy access to all manned areas and living spaces separate from spaces allocated to fight and move functions.
Supportability	A measure of the ease with which a ship can be supported by either maintaining equipment in situ, or easier removal and replacement together with remote diagnostics and condition monitoring.	- Clear vertical removal routes that can be opened easily.
Flexibility	Describes the ability of a complete entity to perform more than one function without any major change. It also implies that the entity has a high availability so that its utility can be exploited.	- General purpose frigate able to contribute effectively against all threats in a battle space.
Re-configurability	Describes the ability of a complete, or part of an, entity to be easily and quickly altered during its life to perform different functions.	- Deck space that can accommodate functionally complete portable containerised systems.
Adaptability	A measure of the ease with which a ship or system design can be changed to produce variants optimised to perform additional or different functions. It is possible that this may be performed on an existing ship at a mid-life rebuild.	- Ship design that has spaces already allocated for in service additions together with in built system capacity to support such additions.

3. DECOMPOSITION

A system decomposition has been undertaken to break the problem down into discrete layers of manageable elements. At the highest level are the system's objectives, at the lowest the physical means of achieving them. Relationships link the highest level characteristics through the layers to measurable parameters at the lowest level. They are shown in Figure 1 and described as follows:

- **Characteristics -** Six system objectives, taken to be the desirable characteristics, commonly, referred to as '-ilities', identified in section 2, Table 3.
- **Benefits -** The definable benefits that are implied by the characteristics are identified explicitly in section 3.1. Some of them appear under more than one characteristic. This is the reason for much of the ambiguity in the use of the characteristics terminology.
- **Design Principles -** In order to gain the benefits there are a set of enabling principles that must be applied in the design of a ship. These are discussed in section 3.2.
- **Design Parameters -** Finally a set of measurable design parameters and specific features are identified which result from the implementation of the design principles. They are intended to assist in the comparison of different designs. The design parameters and features are discussed in section 3.3.

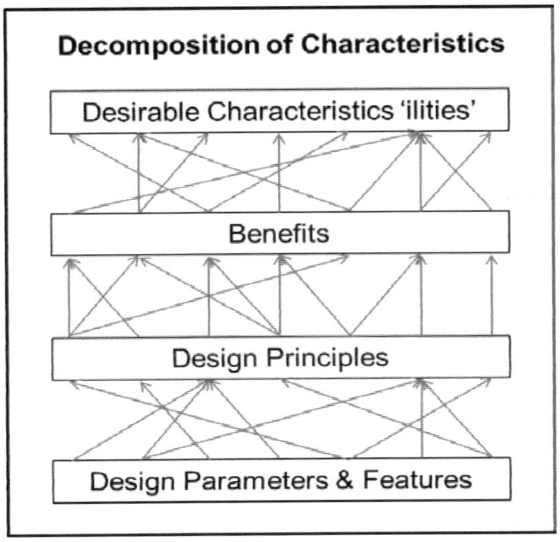

Figure 1: Decomposition of Desirable Characteristics

It should be noted that:-

- **Layers** – The definitions of decomposition layers can be a matter of preference. The breakdown adopted here is intended to illustrate how different perspectives on the problem interact.
- **Elements** – The list of elements in each layer is not exhaustive but is presented as an initial attempt to better define how the implied benefits expected are to be achieved when a ship requirement demands that the '-ilities' are to be provided.

3.1 BENEFITS

The system benefits have been defined as a set of clearly understood abilities described below:

- **Accessibility for Production and Maintenance** - It is very difficult to both install and maintain equipment in a tightly packed compartment. Both individual trade productivity and coordination and sequencing of work across trades are much easier if there is sufficient access space available. The down side is that the ship must be larger with a consequent increase in material, associated labour and running costs required. However the reduction in programme time possible for the lower density ship can potentially compensate for this with subsequent maintenance cost and time reduction benefits through life.
- **Assembly and Setting to Work away from Ship** - In general the manufacture and assembly of a ship will move through several stages, typically

  - Panel line or fabrication shop
  - Module or block assembly hall
  - Building berth
  - Afloat alongside quay

  There is a well known rule of thumb that the later in this sequence that a given task is performed, then the greater will be the labour cost and time required. A factor of two at each stage is often quoted. The earlier assembly work is programmed, then the more efficiently it can be performed, with easier access to appropriate services etc. There are also other cost savings that accrue from moving outfitting tasks earlier as it then becomes easier to smooth out the labour utilisation profiles through the build programme.
- **Standardisation and Repeatability of Work** - If production processes are repeatable then they can become more efficient through both learning and investment in better processes. Larger batch sizes are achieved through commonality of assemblies within and across ships. Again reduced through life costs for both training and maintenance are possible.
- **Rearrangement of Functional Zones** - The concentration of dependent sub systems and spaces into Functional Zones should reduce the changes required to the rest of the ship if subsystems within that zone are changed and as a consequence should reduce the cost of a variant ship. It is highly desirable to be able to supply the same vessel design to multiple customers as this gives benefits

to both supplier and operators. From the supplier's point of view the greater throughput allows a more stable workforce and economies of scale and hence reduced costs. From the customer side reduced prices and risk together with the potential for interoperability with other operators are compelling attractions. However national politics, preferences and operating philosophies together with different procurement timeframes, technology evolution and equipment obsolescence mean that changes to the baseline ship design are always required by repeat and follow on customers. In a conventional ship design, which has extensive system interconnection and distribution throughout the ship, even a small change to some sub systems can have effects throughout the ship leading to a complete re-drawing of the ship and consequently First Of Class (FOC) costs growing to the same level as a completely new ship.

- **Insertion of Additional Functional Zones into Variant Ship** - This describes the ability to add extra blocks into an existing ship design with minimum re-design of the rest of the ship. Some potential variants may require a substantial change in capability that necessitates a lengthening of the ship to accommodate the additional equipment and perhaps increased crew numbers required to operate and support it. Normally this would require a complete redesign of the ship, completely removing any opportunity for cost reduction etc. However if the ship has been designed to allow for changes in length, through the use of Functional Zoning and provision of break points in the interior arrangement and for changes in beam, through isolation of the bulk of the interior outfit from the external shell, then new zones can potentially be inserted into a lengthened ship and changes in beam accommodated without re-modelling and drawing the central outfit core. The need for this is perhaps best illustrated by the fact that several RN combatant classes have undergone major dimension changes from one batch to another e.g. Broad beam Leanders and stretched T42s and later batches of T22s.
- **Re-role of ship by Module Interchange** – Re-role of a ship is possible by swapping out or adding additional equipment rapidly in areas of the ship designed to accept them. In order to reduce the size and hence cost of the ship it may be designed to perform a limited range of roles. This can impose severe operational limitations and in order to alleviate them it is desirable to be able to re-role the ship by rapidly changing modules. The simplest way to achieve this is the provision of an area in the ship that can accommodate containers holding complete systems. These systems will require the provision of power and perhaps fluids and data etc. but their interface with the rest of the ship is likely to be minimal. The containers must be positioned in an area of the ship that enables them to function effectively. For more complex subsystems that are closely integrated with core ship systems, typically weapons system equipment, dedicated optimised locations in the ship with common interfaces to the command system are preferred.
- **Maintenance by Replacement for Complex Equipment** - Ship availability can be increased if key complex equipment can be quickly replaced if they fail. This has been regular practice for prime movers but is less often performed for weapon system elements. In part this is due to the much smaller pool of equipment available, either in the operator's own store and logistic systems or from the Original Equipment Manufacturer (OEM). It is more easily achieved if the equipment is already installed in modules as described in the previous section, however non modular equipment can also be rapidly replaced if provision is made in the ship design. This should be enshrined in a removal route policy that integrates with suitable ship design features such as wide clear access routes, vertical removal routes etc. The benefits are also applicable during build reducing the need to ship vulnerable equipment earlier than strictly necessary because of access limitations when structure is closed up.
- **Rapid Updating of Ship with New Equipment** - It is inevitable that certain equipment will either become obsolete or uneconomic to maintain during the ship life. The more provision is made for easy equipment replacement and removal the better. As such the requirement for this feature is very similar to the maintenance requirement addressed by the previous feature. Ideally the system interface for new equipment will be similar however it is possible that they may have differing system requirements. In such cases the provision of system routes that are clearly separated and easily accessible will make required changes to the interconnection systems easier. It is also possible that completely new equipment will be required providing new capabilities to the ship. Whether this is achievable or not depends on the upgrade margins available in the ship and how they are allocated, i.e. the provision of discrete spaces available For But Not With (FBNW) equipment and excess capacity in connected service systems.

The characteristics influenced by these abilities are shown in Table 4.

Table 4: Characteristics Influenced by Benefits

Feature	Characteristic Affected					
	Producibility	Operability	Supportability	Flexibility	Re-configurability	Adaptability
Accessibility for Production and Maintenance	■	■	■			■
Assembly and Setting to Work away from Ship	■		■		■	■
Standardisation and Repeatability of Work	■				■	■
Rearrangement of Functional Zones	■					■
Insertion of Additional Functional Zones in Variant Ship	■					■
Re-role of Ship by Module Interchange	■	■	■		■	■
Maintenance by Replacement for Complex Equipment	■	■	■	■	■	■
Rapid Updating of Ship with New Equipment	■	■	■	■	■	■

3.2 DESIGN PRINCIPLES

A collection of Design Principles have been identified that have as their objectives the maximisation of the characteristics defined in section 2 Table 3 by enabling the benefits listed in section 3.1 The Principles are classified into groups and arranged in a hierarchy as shown in Table 5, which also shows how the principles are intended to affect the desired top level characteristics. There are of course advantages and disadvantages inherent in implementing these principles in any given design. The main drawback is that many lead to an increase in ship size and hence some costs. However it is important to consider the benefits and indeed cost savings that can be gained by their intelligent use, particularly when they are used together to maximise the potential benefits as shown in Table 6.

There are strong interdependencies between the Design Principles where the implementation of one requires the presence of others. These dependencies are shown in Table 7. This table is read as Row Item depends on Column Items or alternatively Column Item enables Row Items.

Table 5: Design Principles and their Impact on Desired Characteristics

Design Principle Level 1	Design Principle Level 2	Characteristic Affected: Producibility	Operability	Supportability	Flexibility	Re-configurability	Adaptability
Design Geometry	**Design Margins Suitable for Design Intent (DM)** The design margins specified in the Margin Policy must be suitable for the design intent. They must include allowances for both uncertainty and inaccuracies in both known data and parameters not yet defined.	■					■
	Dimension Unit Commonality and Maximisation (DU) This involves using the same basic sizing unit over as wide an area of the ship as possible and also making it as large as is practicable without wasting space. This then facilitates the standardisation of compartments and sub-assemblies with similar functions wherever they are in the ship.	■					■
	Dimension Range Selection (DR) Key overall ship dimensions should be selected so as to be compatible with the full range of potential variant ship dimensions.						■
Space Management	**Space Boundaries for Key Spaces (SB)** Key space or compartment boundaries are defined so that they will remain constant across all possible ship sizes and variants. This means that they must be sized for any likely equipment changes across variants although these should be very limited in most cases, provided system capacities and distribution mechanisms are sized adequately for all variants.	■				■	■
	Space Independence from External Hull or Superstructure (SI) As far as possible the external hull and superstructure shell should be kept separate from a central core of fixed width containing all key compartments. This provides the ability to change the shape of the hull and or superstructure with minimal change to the internal compartments.	■		■		■	■
	Space Standardisation (SS) Wherever compartments serving similar functions are located in the ship they should be identical in size and layout as far as possible. This will reduce drawing requirements through the use of standard layouts. By the same token production efficiencies through repeatability should be achievable.	■				■	■
Functional Management	**Functional Adjacency (FA)** This is the location of several compartments or items of equipment next to one another where there is a net operational or production engineering benefit in collocating them. It is a subset of Functional Zoning.	■	■				■
	Functional Zoning (FZ) This describes the grouping of related operational spaces making up a complete system next to each other in the same part of the ship. The Functional Zone will ideally be completely contained within a single build block or interchangeable module in order to maximise opportunities for early Setting To Work (STW) and testing etc.	■	■			■	■
Interfaces	**Interface Standardisation (IS)** Physical interfaces for power, data and other distribution systems should be managed in order to standardise their form and location thus simplifying manufacture and installation. Wherever available, international standards should be used in preference to national in	■		■		■	■

Design Principle Level 1	Design Principle Level 2	Characteristic Affected					
		Producibility	**Operability**	**Supportability**	**Flexibility**	**Re-configurability**	**Adaptability**
	order to allow easier substitution of baseline equipment for locally sourced alternatives in export variants. The principle also covers modular re-configurable software and consoles etc.	■		■		■	■
	Assembly Interface Standardisation (AI) Where groups of functionally related equipment can be brought together in an assembly on their own sub structure, all the interfaces to the equipment, power, data and fluid etc. should be in the same physical positions allowing interchange with a similar functional assembly. The sub structure interface to the ship should also be standardised wherever possible.	■		■		■	■
	Assembly Self Containment (AS) This covers assemblies that are self-supporting and may be complete sub-systems in themselves including all local system distribution runs to standard interfaces. Such assemblies are commonly referred to as Modules. They may be set to work and tested to a large degree away from the ship and then inserted into the ship as complete items during construction.	■	■	■	■	■	■
Routing	**Routes for Access Wide and Clear of Obstructions (RA)** Access and passageway spaces within the ship are key to ensuring free movement of personnel in a way that optimises operational efficiency and safety. They therefore need to be positioned very carefully in relation to any Functional Zones. In addition to their role in providing personnel access they are also important equipment removal and system distribution routes and as such have a big effect on build efficiency.	■	■	■	■		■
Routing	**Routes for Removal Clear of Outfit (RR)** Routes for removal and replacement of equipment should not be obstructed. Equipment should not intrude into passageways. Vertical removal routes should be clear of system distribution routes, especially power and data distribution cabling, and should not pass through high outfit density compartments. In terms of the difficulty involved if this is not possible, cabling poses the most problems, then pipes and finally ventilation ducting.	■		■	■	■	
Routing	**Routes for Systems Separate from Compartments (RS)** Wherever possible it is good practice to avoid main system routes passing through outfitted compartments. Not only does this make design changes and build much easier but it avoids the risks associated with equipment damage from fluid leaks etc. and maximises the headroom available for equipment or deck-head linings in living areas.	■		■		■	■
Routing	**Routes for Systems in Layers Away from Each Other (RL)** Routes for different systems should be kept away from each other, either by lateral separation in different passages or trunks or more usually by allocating different levels in the deck head space. The use of different levels is of course essential where systems cross each other. Separation of systems into different levels considerably simplifies planning, sequencing and installation work.	■		■		■	■

Table 6: Benefits Derived from Combining Design Principles

Design Principle Level 1	Design Principle Level 2	Benefits							
		Accessibility for Production and Maintenance	Assembly and Setting to Work away from Ship	Standardisation and Repeatability of Work	Rearrangement of Functional Zones	Insertion of Additional Functional Zones in Variant Ship	Re-role of Ship by Module Interchange	Maintenance by Replacement for Complex Equipment	Rapid Updating of Ship with New Equipment
Design Geometry	DM	■			■	■			■
	DU			■		■	■		
	DR					■			■
Space Management	SB			■	■	■			■
	SI	■				■			
	SS			■					■
Functional Management	FA	■	■		■	■			
	FZ	■	■		■	■			
Interfaces	IS			■	■	■	■	■	■
	AI			■			■	■	■
	AS		■				■	■	
Routing	RA	■						■	■
	RR	■					■	■	■
	RS	■			■	■	■	■	■
	RL	■		■					■

Table 7: Interdependencies between Design Principles

Design Principle Level 1	Design Principle Level 2	Row Dependent on Column/Column Enables Row														
		DM	DU	DR	SB	SI	SS	FA	FZ	IS	AI	AS	RA	RR	RS	RL
Design Geometry	DM	░														
	DU		░													
	DR		■	░												
Space Management	SB		■	■	░											
	SI		■	■	■	░										
	SS		■	■	■	■	░									
Functional Management	FA		■		■			░								
	FZ			■	■			■	░							
Interfaces	IS									░						
	AI		■	■	■	■	■			■	░					
	AS				■	■		■	■	■	■	░				
Routing	RA		■		■								░			
	RR				■									░		
	RS		■				■						■		░	
	RL												■		■	░

3.3 DESIGN PARAMETERS AND FEATURES

In order to simplify the assessment of ship impact a series of design parameters and features were identified that are directly influenced by the application of the design principles. These parameters and the design principles that they are related to are shown in Table 8.

Table 8: Design Parameters and Features and their Relationship with Design Principles

Design Parameter or Feature	Dependent on Design Principles														
	DM	DU	DR	SB	SI	SS	FA	FZ	IS	AI	AS	RA	RR	RS	RL
Deck height															■
Seating weight to total structural weight ratio											■				
Access and passage way area ratio					■							■	■	■	■
Accommodation area per man				■		■					■				
Group 4 Command and control weight to space ratio	■	■	■	■											
Group 5 Auxiliary machinery weight to space ratio	■	■	■	■		■				■	■				
Group 7 Weapon weight to space ratio	■	■	■	■		■				■	■				
Dedicated system galleries												■	■	■	■
Mission bay			■				■	■	■	■	■				
Payload modularity				■		■	■	■	■	■	■	■			
Growth margins	■		■	■									■	■	
For But Not With (FBNW) future equipment	■		■	■					■	■	■		■	■	

Analysis of existing designs allows these parameters to be quantified and compared. An example of such a comparison is shown in Figure 2 for Group 4, 5 and 7 equipment densities representing Command, Control, Communications, Computers and Intelligence (C4I), auxiliary machinery and weapons systems respectively. The use of a ship concept programme such as BAESHIP™ allows the effect on overall ship size of varying these parameters, both individually and collectively to be studied (2). An example of the effect of varying auxiliary equipment density on a baseline concept frigate design is shown in Figure 3. In this plot selected parameters relevant to the overall cost of the ship are shown. These parameters are shown to represent factors that can act to either increase or decrease different cost elements. Although changes in the driving parameters caused by the application of design principles will generally increase ship size, this does not necessarily mean that the overall cost of the ship increases. It is important to recognise that overall cost is made up of many cost elements. Key among these is overall programme time. The application of modularity and increased access during build are both well known enablers to reducing programme time and hence cost. They are also significant contributors to maintainability and through life costs.

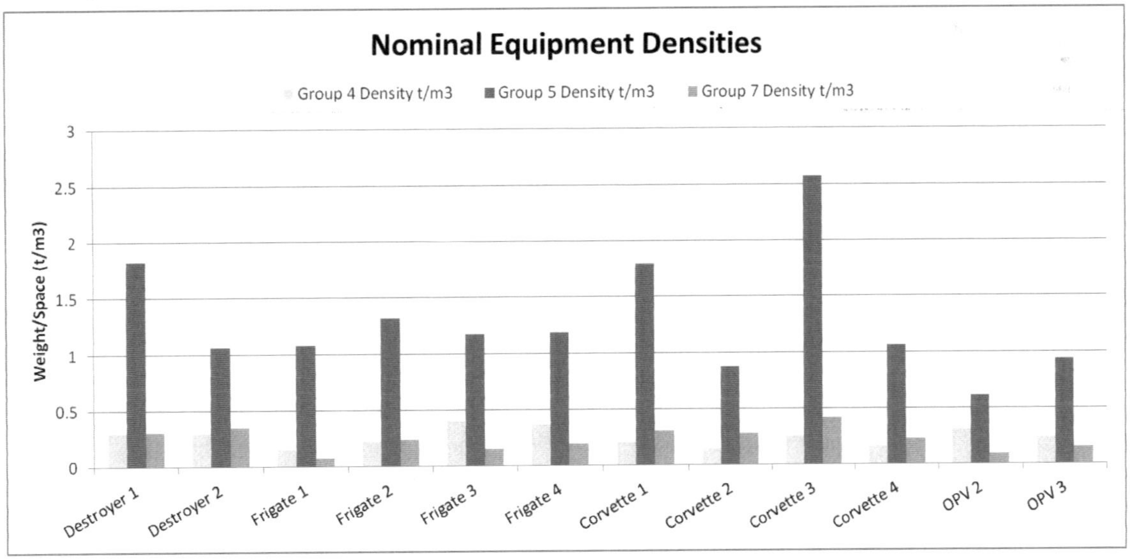

Figure 2: Equipment Densities for C4I, Auxiliary and Combat Systems Equipment

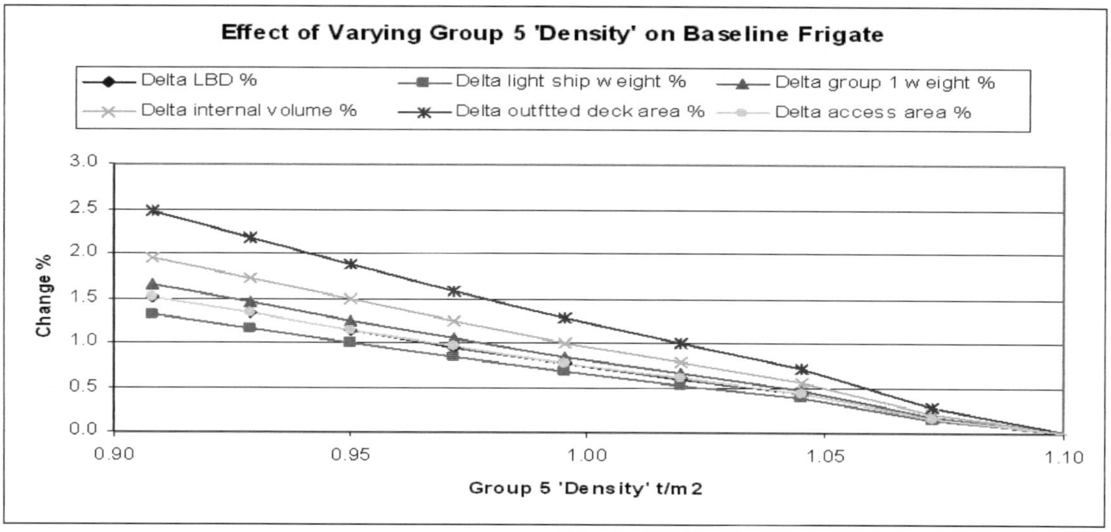

Figure 3: Effect of varying Auxiliary Equipment Outfit Density on Selected Ship Cost Parameters for Baseline Concept Frigate

BAESHIP™ was used to generate a baseline frigate model based on current design practice, which uses some of the principles, and an alternative concept applying most of the principles described. The differences in key parameters between the baseline and modified designs are shown in Table 9.

Table 9: Comparison between Baseline and Modified Frigate Concepts

Category	Parameter	δ%
Ship particulars	Length waterline m	12.6
	Beam waterline m	1.4
	Displacement t	10.3
Changed Design Parameters	Deck height	0
	Seating wt/Group 1 wt	9.5
	Access area/Ref outfit area	44.6
	Accommodation area/man	0
	C4I equipment t/m2	0.6
	Auxiliary equipment t/m2	-13.3
	Weapon equipment t/m2	-8.8
	Dedicated system galleries	Added
	Mission bay fitted	Neither
	Payload modularity	Added
	Growth margin	Same
	FBNW future equipment	Neither

The data generated together with the use of the BAESHIP™ concept tool thus allow a preliminary estimate of the effect of applying different design philosophies to a concept to be estimated. Cost estimates can be obtained for the new concepts by using cost estimating relationships derived from the design parameters and known cost elements of existing ships. These must then be modified by estimates of productivity improvements for relevant cost elements enabled by the changes in design parameters.

It is also possible to construct a value assessment model using the system modelling programme BAEFASIP™ (2) based on the structure shown in Figure 1 and the relationships described above. This can then be used to compare the baseline and modified frigate designs in terms of the desirable characteristics using the design parameter values as input. The actual values for the characteristics will depend heavily on any preference weightings applied to the linkages in the model. As an example a simple comparison between the baseline and modified frigate concepts using Value Theory type preference swings of 0-1 and equal weights throughout is shown in Figure 4.

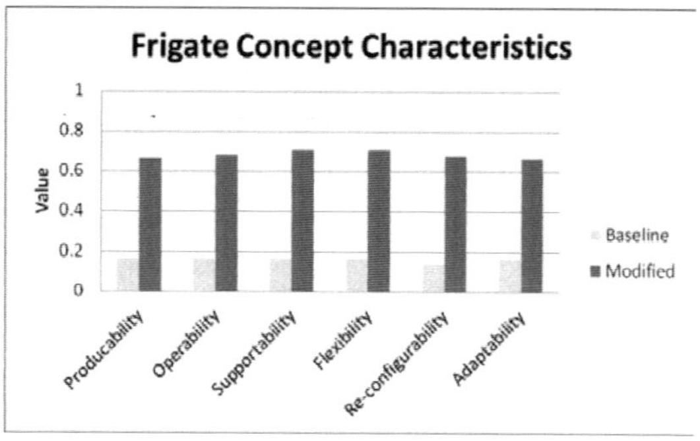

Figure 4: Comparison of Desirable Characteristics for Baseline and Modified Frigate Concepts

Such numerical comparisons become useful when the system structure is used to generate a shared understanding of the entity interrelationships and so reach agreement on preferences thus enabling a full range of concepts to be compared on a rational basis.

4. COMBAT SYSTEM MODULARITY

The next stage of the work was to consider in more detail the potential benefits and implications of the use of combat system modules by the RN. Such systems have of course been used for many years, the best known being the German MEKO® and Danish StanFlex™ systems. A review of these systems indicated that the benefits that have been achieved are substantial but were not always those that were originally intended.

4.1 ADVANTAGES

The advantages of payload modules were identified as

- **Producibility** - A major benefit of self-contained weapon modules is that they can be assembled, set to work and tested away from the ship. This both reduces direct costs and allows a shortened build time scale, which further reduces total cost.
- **Re-configurability** - In service it is possible to reconfigure a ship for alternative roles by changing certain modules. This may take a few days as system safety checks and integration testing will still be required each time a module is installed (these can dominate the times).
- **Rationalisation** - Weapons are expensive and if they are not required for all the roles that a ship may be called on to perform then the fleet holding may be reduced from the usual one per ship plus reference. Ships are only fitted with the equipment when they are assigned to the roles that require it. This of course compromises flexibility to some extent but has the benefit of substantial cost savings.
- **Re-usability** - The use of modules allows weapons to be transferred from older to replacement ships very easily and cheaply. Weapon equipment service life and upgrades can thus be decoupled from those of the ships themselves leading to much smoother budget profiles and simpler and cheaper acquisition programmes.
- **Availability** - Total system availability is increased as defective weapons can be immediately replaced by swapping modules, assuming suitable supporting infrastructure is in place.
- **Development -** It has been shown repeatedly that it is impossible to link the development of weapon systems to the design and build of a new class of ship without increasing the costs and timescales of both ship and weapon. The use of modules allows the decoupling of weapon system development from ship replacement. This may be particularly significant with the development of self-defence directed energy weapons and off board systems, both of which are likely to evolve rapidly over the next few years. Incremental acquisition and in service upgrade of ships weapon systems becomes much easier and can be decoupled from both major platform procurements and refits.

In the light of these conclusions an assessment was made of how to maximise the potential benefits for the RN. The result was the development of a proposed strategy based around 5 key module elements that would maximise the overall benefits to the fleet (3). The strategy elements and their advantages are summarised in Table10.

Table 10: Proposed Weapon Modularity Strategy – Elements and Advantages

Module			Advantages					
Class	**Type**	**Application**	**Producability**	**Re-configurability**	**Rationalisation**	**Re-usability**	**Availability**	**Development**
(1) Large VLS system	(1A) Multiple deck height cellular	AAW missiles						
		Land attack missiles						
		SSM						
		ASW missiles						
	(1B) Half size ISO	Missile fire control units						
(2) Self Defence Module System	(2A) Small footprint deck penetrating single deck height self-contained modules with deck mounted equipment	Sensors, directors						
	(2B) Medium footprint deck penetrating single	Above water decoys						
		SCG systems						

Module			Advantages					
Class	**Type**	**Application**	**Producability**	**Re-configurability**	**Rationalisation**	**Re-usability**	**Availability**	**Development**
	deck height self-contained modules with deck mounted equipment (possibly with underdeck magazines)	CIWS	■	■	■	■	■	■
		Short range VLS missiles	■	■	■	■	■	■
		Short range trainable missiles	■	■	■	■	■	■
		SSM (overlapping upper deck footprint)	■	■	■	■	■	■
	(2C) Ready use magazines	Plug in deck magazines		■	■	■	■	■
		Modular racking and spray systems for in built magazines		■	■	■	■	■
(3) ISO Standard Compliant Containers	(3A) COTS full ISO standard envelope compliant	Accommodation		■	■	■	■	■
		Hospitals/ medical		■	■	■	■	■
		Workshops/ maintenance		■	■	■	■	■
		Offices		■	■	■	■	■
		Stores		■	■	■	■	■
		Equipment- generators, RO plant etc.		■	■	■	■	■
	(3B) MOTS full ISO standard envelope compliant	UAV L & R		■	■	■	■	■
		UXV C2	■	■	■	■	■	■
		MCM C4	■	■	■	■	■	■
		Satcom, specialised IT and Intelligence	■	■	■	■	■	■
	(3C) MOTS ISO standard footprint compliant, varying height	Small boat/ USV/ AUV and side L & R system		■	■	■	■	■
		Small boat/ USV/ AUV and stern L & R system		■	■	■	■	■
(4) Dedicated Towed Systems	(4A) Small (possibly ISO compliant)	Under water decoys	■	■	■	■	■	■
		Side scan sonars	■	■	■	■	■	■
		Small MCM sweeps	■	■	■	■	■	■
		CTD profilers	■	■	■	■	■	■
	(4B) Large Heavily platform integrated	Large sonars	■	■	■	■	■	■
		Large MCM sweeps	■	■	■	■	■	■
(5) Large Boat/ USV/ AUV Handling System Interface	(5A) Vehicle and cradle system for side L & R system	Boats		■	■	■	■	■
		USVs		■	■	■	■	■
		AUVs		■	■	■	■	■
	(5B) Vehicle and cradle system for ramp L & R system	Boats		■	■	■	■	■
		USVs		■	■	■	■	■
		AUVs		■	■	■	■	■

Figure 5 and Figure 6 show how these modules can be used on a surface combatant and an RFA.

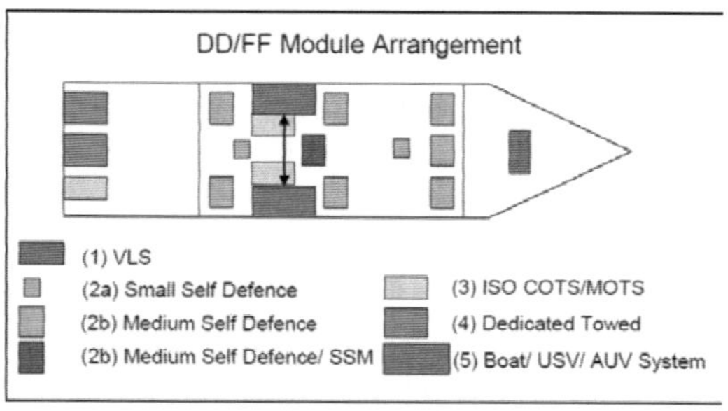

Figure 5: Use of Modules on a Surface Combatant

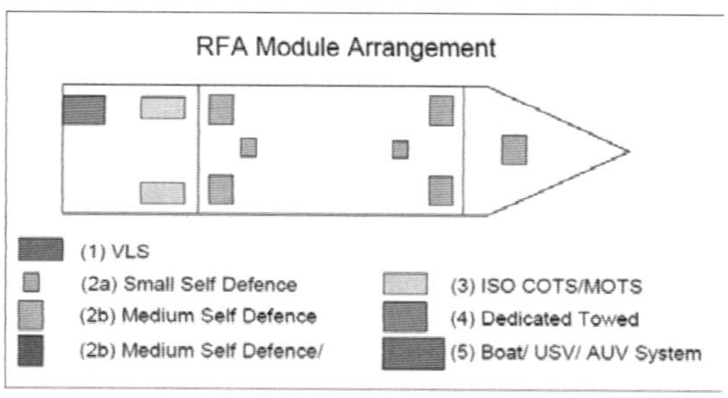

Figure 6: Use of Modules on an RFA

5. ALTERNATIVE FUTURE CONCEPTS

Given the obvious advantages of adopting a modular combat system approach the next stage was to generate a range of ship surface combatant concepts that included both adaptations of existing designs without modules and new concepts with them. The concepts with modules attempted to exploit the benefits offered by both role re-configuration and the use of evolving off board systems. The new designs were generated using the BAESHIP™ software and then developed to the stage that they could be costed with a reasonable degree of confidence. A plot of displacement against waterline length for the concepts and a range of recent RN and contemporary European vessels is shown in Figure 7. It is noticeable that there has been a gradual decrease in ship's circular M (Waterline length / Displaced volume$^{1/3}$) over recent years. There are probably a variety of reasons for this including higher top weights required for more sensors, higher deck heights and heavier structures driven by producibility concerns.

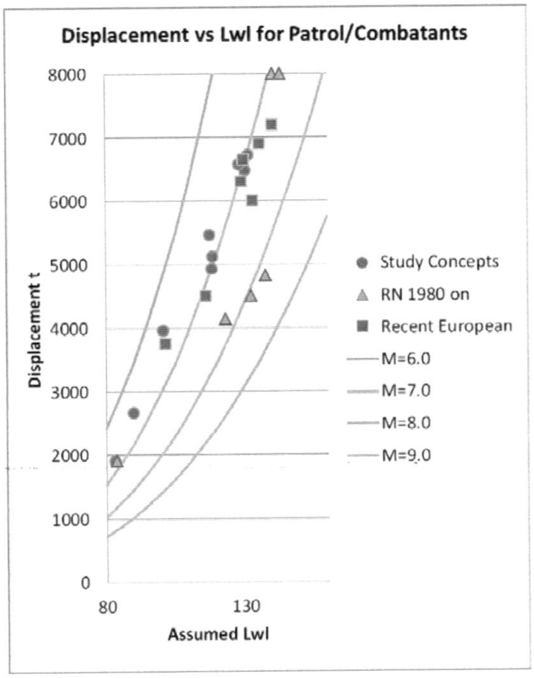

Figure 7: Displacement vs Waterline Length for Existing and New Concept Surface Combatants

The concepts along with in service and planned RN combatants were then assessed by subject matter experts against the capabilities required by current operational requirements. A capability assessment plot for concept ST6, an escort frigate focussed on above water warfare, is shown in Figure 8.

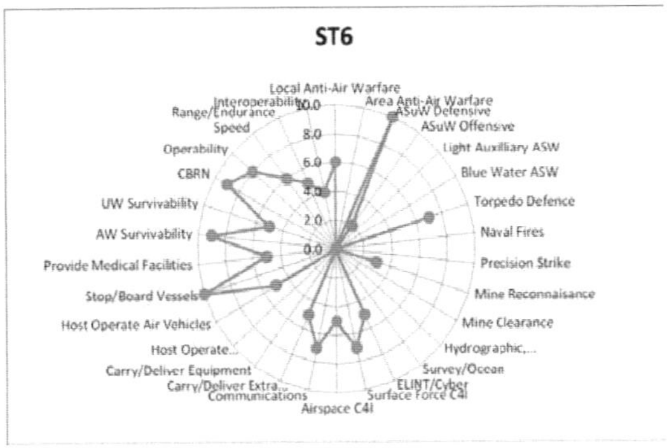

Figure 8: Capability Assessment for Concept ST6

6. FLEET STRUCTURE OPTIMISATION

The previous studies had generated cost and military capability effectiveness data for a range of existing and new concept designs. The assessed capabilities were then mapped onto a set of military operations drawn from the then current defence policy. These operations were then mapped onto sets that could be expected to occur concurrently. The largest single concurrency demand for each ship type then resulted in the minimum fleet size. These mappings enabled a BAEFASIP™ model to be constructed that could be used to produce a Pareto front of optimum cost effectiveness fleet mixes (4). The model was constructed from multiple interconnected layers and optimised the allocation of ships to each operation taking account of

- Ship capabilities and capacity for deploying off board systems (the same mechanism can also be used to account for module usage).
- Numbers of ships with particular capabilities required by operations.
- Numbers of ships with capabilities above a critical level required by operations.
- Numbers of off board systems required by operations.
- Operations required by each concurrency.
- Force generation factors for each ship type.
- Manning and associated generation factors for each ship type.
- Limits on numbers of a ship type (e.g. for T45).

It performed a multi objective optimisation on five parameters

- **Costs** – Acquisition costs comprising FOC and unit production costs (minimise).
- **Total benefits** – calculated from the scores of the allocated ships for the required capabilities for each operation (maximise).
- **No. of Concurrencies** – count of the number of required concurrencies that have all their requirements met in terms of required minimum

capability scores and numbers of units of capability (maximise).

- **Total manning** – considered as a major through life cost driver (minimise).
- **Number of classes** – considered as a significant through life cost driver (minimise).

The result of the optimisation was a Pareto Front representing the trade-offs between the optimised variables. The optimum front is shown in Figure 9, Figure 10 and Figure 11 for three variables at a time.

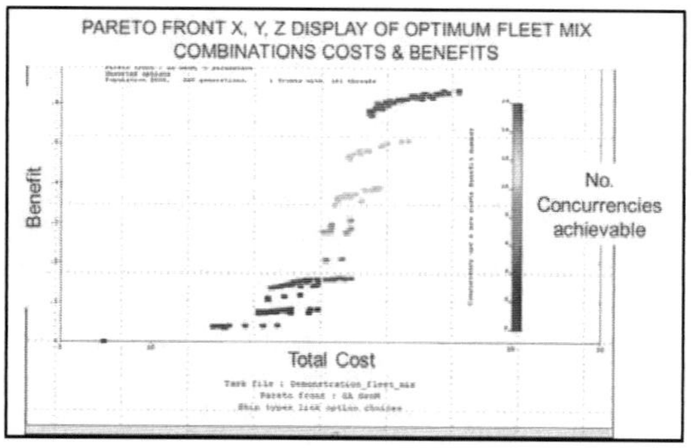

Figure 9: Benefit vs Cost – No. Concurrencies

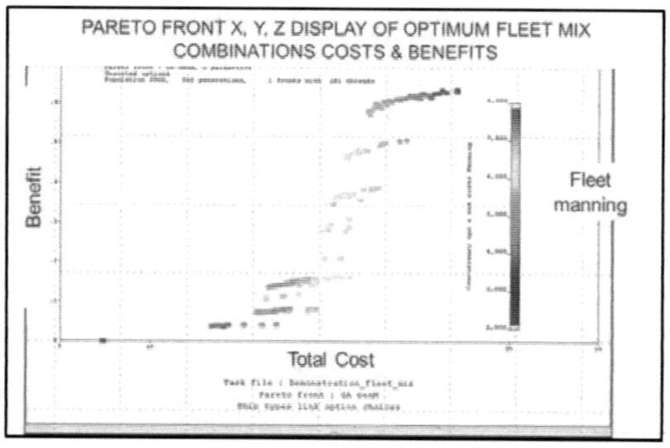

Figure 10: Benefit vs Cost - Manning

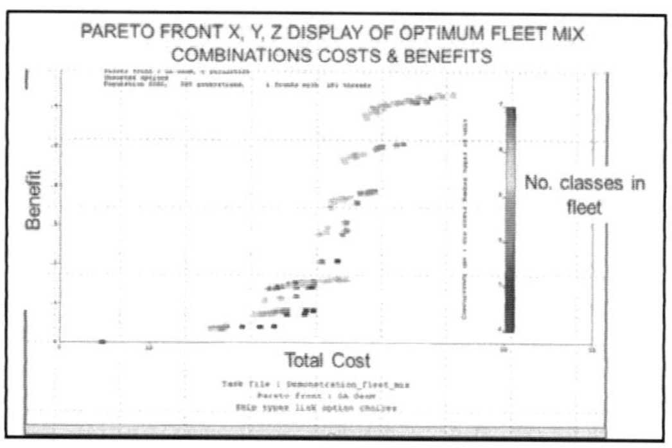

Figure 11: Benefit vs Cost - No. Classes

The numbers of each ship type making up the optimum fleet mixes that achieved all the required number of concurrencies are plotted in increasing fleet cost order in Figure 12.

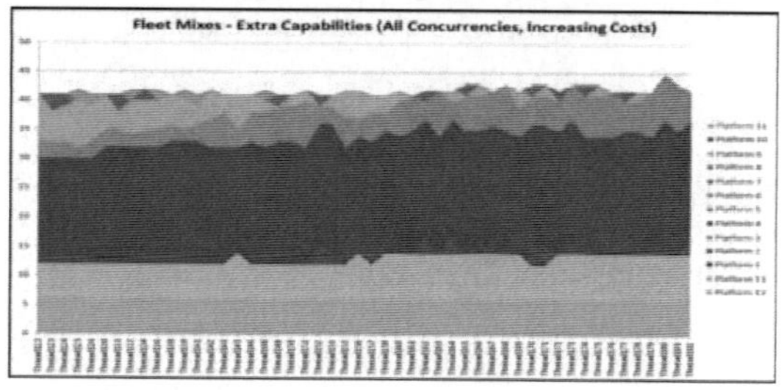

Figure 12: Optimum Fleet Mixes in Increasing Cost Order

The results showed that the most favoured ships were larger ones able to deploy a range of capabilities particularly boats and off board systems.

7. DISCUSSION

The paper has offered but one approach to clarifying the benefits and implications of Adaptability, Flexibility and Modularity. Other interpretations are possible but whatever approach is used the complex interrelationships between them must be clearly understood if the expected benefits are to be obtained. The studies and paper have been limited in scope and there are evidently further interactions to be considered for example:

- Operability could be extended to include survivability and as such could be negatively influenced by functional zoning. Such considerations are perhaps best considered in separate studies and then brought together in a full trade space cost benefit analysis.
- Additional Design Parameters and Features that are cost drivers could be considered.
- Further analysis on the realisable benefits is needed, for example the impact of logistics and availability of appropriate combat system modules in specific scenarios.

8. CONCLUSIONS

The terms Adaptability, Flexibility and Modularity are often used ambiguously and are seen as solutions to the affordability crises facing navies today. The paper has shown that the intentions are better described by six 'ilities' referred to as Desirable Characteristics each with clearly defined meanings,

- Producibility
- Operability
- Supportability
- Flexibility
- Re-configurability
- Adaptability

Further studies have established methodologies and a toolset that can both assist in exploiting the potential benefits during the development of new concepts and optimising the cost effectiveness of future fleet structures. More specifically the following conclusions can be drawn

8.1 DESIGN INFLUENCES

The work has defined and clarified the opportunities and implications of exploiting the expected advantages of Adaptability, Flexibility and Modularity and illustrated both how they can be achieved and how the implications on ship design, producability and cost can be estimated. It can be stated that

- The benefits are realised as a set of abilities
  - Accessibility for Production and Maintenance.
  - Assembly and Setting to Work away from Ship.
  - Standardisation and Repeatability of Work.
  - Rearrangement of specific Functional Zones without affecting the rest of the ship.
  - Production of variant ships by changing the ship dimensions and inserting additional functional zones without affecting the bulk of the rest of the ship.
  - Re-role of a ship in service by swapping interchangeable modules.
  - Easy removal and maintenance of complex equipment through the provision of better access routes.
  - Updating of a ship quickly with new equipment by the provision of better access to that equipment.
- The application of a coherent set of Design Principles can
  - Improve the Producability of a vessel by enabling the production and in some cases STW of complete modules away from the berth.
  - Improve the Operability of a ship by keeping operational spaces away from living spaces and improving access, transit and evacuation throughout the ship.
  - Improve the Supportability of a ship by providing better access to equipment requiring maintenance and in some cases simplifying repair by replacement.
  - Enable particular areas of a ship to be Re-configured in service to provide new capabilities without affecting the rest of the ship keeping related compartments together in discrete Functional Zones.
  - Enable the production of ship designs that are Adaptable in that variants with different capabilities can be produced without excessive re-design costs through the use of re-usable discrete zones and the ability to change the ship hull shape without affecting the layout of the bulk of the outfitted compartments.
- The Design Principles are strongly interdependent and cannot be applied in isolation from each other if the full benefits are to be gained.
- The effect of applying the Design Principles on the size, cost and work content of the ship can be assessed by relating them to Design Parameters describing quantifiable features of the ship design.
- Although the effect of varying the Driving Parameters on the ship can be quantified the actual degree to which they should be varied in order to gain the full benefits of Producability, Adaptability etc. are currently more difficult to estimate and are a matter of judgement.

8.2 DESIGN STUDIES AND FLEET MIX OPTIMISATION

The results generated from the design and optimisation studies should be considered preliminary as the work to date has inevitably opened up many questions and possibilities that will need to be explored more fully. However it has already been shown that

- The high level cost effectiveness of conventional and modular ships can be examined at a fleet level using the BAEFASIP™ methodology.
- Initial results indicate that
  - The right sub-sets of the total range of capabilities in larger multi-purpose vessels are more cost effective than smaller role specific or larger fully flexible ships. In the initial results the most promising multi-purpose vessels were those that included general escort, HADRO and MCM capabilities but excluded the more expensive area AAW and ASW roles assigned to the T45 and T26.
  - There is considerable value in incorporating comprehensive facilities for boats and off board systems.

8.3 OPPORTUNITIES

The work has shown that it is now possible to explore and hence make better informed decisions on the measures that can be taken to

- Improve a ship design's potential Adaptability and Producability at the earliest stages of design using the BAESHIP™ concept tool.
- Understand the trade offs between Flexibility and Affordability and how Operability and Supportability can be improved by Re-configurability.
- Optimise the high level cost effectiveness of fleet structures taking into account the issues of combat system Modularity and Re-configurability using the BAEFASIP™ methodology.

9. ACKNOWLEDGEMENTS

The authors thank BAE Systems Maritime – Naval Ships and DSTL for permission to publish this paper and would like to acknowledge the help of their colleagues in its preparation.

10. DISCLAIMER

The views expressed in this paper are those of the authors and do not necessarily represent those of BAE Systems Maritime – Naval Ships or DSTL.

11. REFERENCES

1. COURTS M, FIGG K; 'Adaptability and Modularity for RN Platforms'; *BAE Report for DSTL AS-1 Aug 2012.*

2. COURTS M, BRITTAIN R, LAMBLE A, OSBORNE N, 'Warship Trade Space Exploration: Challenges and Approaches'; *RINA Systems Engineering Conference, London; Mar 2012.*

3. COURTS M, 'Proposed UK Way Ahead on Modularity', *DSTL Report DSTL/TR75915; Jan 2014.*

4. COURTS M, BRETT C, 'Combatant Design and Fleet Mix Assessment and Optimisation using BAEFASIP', *34 ISMOR, Royal Holloway, University of London UK, 2017.*

12. AUTHORS BIOGRAPHIES

Malcolm Courts is an Engineering Manager with involvement in concept development and product and design process improvement.

Chris Broadbent is a Dstl Fellow in Platform Systems Above Water Group.

Bringing Medical Facilities to the Forefront of Naval Ship Design, Focusing on Mission Adaptability and the Integration of Modular Medical Facilities

P.Morales and **H.Schweidler**, Babcock International Group, Marine & Technology

SUMMARY

This paper aims to clarify the design requirements and applicable standards for a Medical Role 2 Afloat capability and understand how these facilities could be augmented or scaled, with the use of modular facilities, both for capability and capacity. The research covers the standards and guidance available, their shortfalls and with the use of two case studies, how the standards and guidance can be applied. This paper also investigates the applicability of using custom 20 foot ISO Container sized modules to enhance an existing medical facility on board a vessel and identify what design characteristics would be most applicable.

This paper has been written with the support of the Royal Navy Medical Service, MOD DE&S Commercially Supported Shipping and Marshall Aerospace and Defence Group.

ACRONYMS

CSS	Commercially Supported Shipping
CVS	Carrier
CBRN	Chemical Biological Radiological Nuclear
DD	Destroyer
DCR	Damage Control Resuscitation
DCS	Damage Control Surgery
DE&S	Defence Equipment & Support
DEFSTAN	Defence Standard
DFID	Department For International Development
DEFSTAN	Defence Standard
FF	Frigate
HADRO	Humanitarian And Disaster Relief Operation
HDU	High Dependency Unit
ICU	Intensive Care Unit
ILO	International Labour Organisation
IMO	International Maritime Organisation
JRNMS	Journal of the Royal Naval Medical Service
Lit'M	Littoral Manoeuvre
LPH	Landing Platform Helicopter
LPD	Landing Platform Dock
LRNSR	Lloyds Register Naval Ship Rules
LSDA	Landing Ship Dock Auxiliary
RNMS	UK Royal Navy Medical Service
NATO	North Atlantic Treaty Organisation
MCA	Maritime Coastguard Agency
MCMV	Mine Countermeasure Vessel
MOD	Ministry Of Defence
MTF	Medical Treatment Facilities
NEO	Non-combatant Evacuation Operation
NDP	Naval Design Partnering
ODP	Operating Department Practitioner
OPCP	Operational Patient Care Pathway
OSV	Ocean Survey Vessel
OT	Operating Theatre
POW	Point Of Wounding
RESUS	Resuscitation
R(#)	Medical Role (number designator)
R(#)A	Medical Role (number) Afloat
R(#)B	Medical Role (number) Basic
R(#)E	Medical Role (number) Enhanced
RFA	Royal Fleet Auxiliary
RN	Royal Navy
SME	Subject Matter Expert
STRATEVAC	Strategic Evacuation (to a Role 4 MTF)
T1	Immediate Treatment Casualty Group
TACEVAC	Tactical Evacuation (to a Role 3 MTF)
UN	United Nations

1. INTRODUCTION

The provision of dedicated medical facilities on naval warships and auxiliaries has long been established; however the design of these facilities rarely makes the top list of priorities for the naval architect and is often overlooked in its importance due to emphasis on other platform capabilities.

The Role 2 Afloat (R2A) medical facility is of key strategic importance, as it provides readily deployable basic facilities for resuscitation and life-saving treatment and surgery, prior to evacuation to higher Medical Role hospital facilities (see Table 1 for further Medical Role definitions). However, the medical support provided by platforms with R2A facilities is typically only a secondary role for these platforms. This lower priority is exacerbated by the increasingly multi-role nature of modern naval platforms [1], whereby the vast number of key roles now supported by a single platform creates considerable competition for prime space.

At the same time, medical facilities have traditionally been designed to provide capability in times of conflict, however there has been a recent increase in Humanitarian and Disaster Relief Operations (HADRO), as illustrated by the increase in spending on humanitarian aid [2], supported by the UK RN and other navies.

Over the past few years naval resources have been stretched, with events such as the Mediterranean migrant crisis, evacuation of civilians from overseas conflict (e.g.

Libya) and aid to the Caribbean in the aftermath of Hurricane Irma, to the extent that platform medical facilities may have been used outside of their original design intent.
This implies that new approaches to bolster a navy's capability to support HADRO will be welcomed. The ability to adequately prepare platforms for these roles, which require different facilities from war fighting missions, is paramount to their success.

This paper aims to assist the ship designer by summarising the key rules and regulations governing the design of R2A Medical Treatment Facilities (MTFs) and providing further user supplied guidance where information is lacking. It uses two multi-role vessel concept designs of different size and capability to illustrate how the functionality of the medical capability can be improved through optimal layout of MTFs and vessel arrangement.
The paper also explores how the provision of modular medical units (in regular use for field hospitals) can be exploited within the confines of a naval vessel.

The overall aim is to provide a basis for an adaptable and flexible shipboard MTF, capable of addressing the ongoing need to support HADRO without unduly compromising the other aspects of a warship.

2. INVESTIGATION BACKGROUND & APPROACH

"Space is often more of an issue, as the medical lay-down doesn't always feature heavily in the mind of the naval architects." [3]

This quote was extracted from an article within the Journal of the Royal Naval Medical Service (JRNMS). Similar JRNMS articles indicate that whilst medical facilities are not always ideal, Medical staff will make it work somehow. Clearly, this is far from ideal and will be to the detriment to the efficiency of the workflow.

There are, however, a number of key issues, which hinder the naval architect providing the optimal design of an MTF:

- The increasingly multi-role nature of platforms [1] means that, on a Role 2A vessel, where medical support is often a secondary role, there is huge competition for space and prime locations with other key role requirements.
- NATO guidance for each role provides little detail on what should be provided on the ship in terms of areas and layout. They have been written to enable the host nation to provide tailored guidance to the designer.
- DEFSTAN guidance on Medical Facilities do not align fully with current UK RN Medical Doctrine and are currently outdated and under review.
- Guidance on staffing numbers, ratios, total casualty beds and type is not readily available to the designer.
- The MTFs of existing platforms differ significantly in design and layout (for the same class) and facilities are often not used in the way that was intended when they were designed, making it hard to gain guidance from previous designs.
- Due to the slow procurement times associated with naval ship design and build, the MTFs provided are often out dated by the time they reach service.
- There is a rise in the need for naval platforms to support HADRO and it is unclear how to best support this within the confines of the ship.
- Regardless of whether in date standards are available, it is still essential to establish a dialogue with the end user, e.g. the UK RNMS, in conjunction with the Authority.

The design and layout of the MTF within the ship, as well as the supporting systems, accommodation and facilities, will dictate the efficiency of the workflow, the ease in which the medical staff can carry out their work and ultimately the wellbeing of the military casualties.

An understanding of the workflow and how facilities are used is always best sought from the end user. It is the key to defining spatial and arrangement requirements within a ship and best established at the start of the design process (if space is not assigned at a concept stage it is very difficult to obtain it in the later design stages).

In order to address these issues a workshop with the RNMS was organised at the Institute of Naval Medicine in Gosport.

The workshop covered the following:

- Available guidance and standards;
- Workflow and compartment layout within MTFs;
- Staff ratios and accommodation requirements;
- Review of existing vessel MTFs;
- Review of two concept MTFs;
- The layout of a typical future R2A MTF and how it may be designed in a future proof manner;
- Modular medical facilities, the pros and cons and how they could be used on naval platforms;
- Design workshop using two concept designs to establish what additional facilities would be most beneficial for what mission type and how modular units could be arranged within the ship.

3. DOCTRINES, STANDARDS & REGULATIONS

3.1 MEDICAL CAPABILITY AND ROLES

Deployed Hospital Care is a vital part of the Defence Medical Services and is delivered on a joint basis across the armed forces, with the aim of returning the sick and injured back to duty as quickly as possible. This care is delivered through a number of medical capabilities linked to the Operational Patient Care Pathway (OPCP) [2 &5] which fall within the NATO Medical Role concept [5] – see Table 1 below.

NATO ROLE DEFINITIONS	
ROLE 1	**Primary Healthcare:** **GP Service (and Dental depending on size/class of vessel)** Pre Hospital Care - Triage, First Aid, Damage Control Resuscitation (DCR)
ROLE 2 BASIC - 2B	**Deployed Hospital Care (fwd): Damage Control Surgery (DCS)** Has all Core NATO Medical Modules Also includes: Role 1 Facilities Pre Hospital Care - Triage, First Aid, DCR (TACEVAC to ROLE 3 or STRATEVAC to ROLE4)
ROLE 2 ENHANCED-2E (2 AFLOAT - 2A)	**Deployed Hospital Care (fwd): DCS and limited General, Vascular and Orthopaedic Surgery** Has all Core NATO Medical Modules and some enhanced modules - blood banks/ labs etc. & more capacity Also includes: Role 1 Facilities Pre Hospital Care - Triage, First Aid, DCR (TACEVAC to ROLE 3 or STRATEVAC to ROLE4)
ROLE 3	**Deployed Hospital Care (rear): General & Specialised surgery** Has all Core NATO Medical Modules and most enhanced modules. Also includes: Role 1 Facilities Pre Hospital Care - Triage, First Aid, DCR Role 2 Facilities (STRATEVAC to ROLE 4)
ROLE 4	**Full Hospital Care**- Always land based in the host nation (UK - Queen Elizabeth, Selly Oak, UK)
Notes: The defining facilities for the Role is in **bold**. However, vessels with higher Role MTFs also provide the facilities of lower Role MTFs	

Table 1: Definition of NATO Medical Roles – derived from [5]

The OPCP, when used in conjunction with the medical 10, 1, 2, (2), +2 planning guidelines, identifies the time scales between the point of wounding (POW) and transfer to a Role 4 facility. (Note that the critical time scales varies slightly between NATO nations.)

- POW +10 mins – Enhanced First Aid,
- POW + 1 hr - Enhanced Field Care/DCR, (R1)
- POW + 2 hrs - Deployed Hospital Care, DCR/DCS (R2),
- (2hrs) – typical duration of surgery,
- +2 hrs - maximum time for tactical or strategic evacuation (TACEVAC or STRATEVAC) after DCS to in theatre surgery/ enhanced diagnostics (to R3 or R4).

Critically the R2A platform allows for deployed hospital care to be brought closer to the POW, helping to provide the vital care required within the medical timelines allowing for greater operational agility prior to onward movement to Role 3 or Role 4 care.

3.2 STANDARDS

As is the case for most ship board facilities, there is no single point of truth for the design of MTFs. In a review of available standards and regulations the following were deemed of key importance to the design and layout of MTFs on board naval ships:

3.2 (a) NATO Standards

- ANEP 77: Naval Ship Code [6]
- AJP-4.10 – Allied Joint Doctrine for Medical Support - Edition B Version 1 + UK national elements MAY 2015 [5]

The Naval Ship Code – ANEP 77 does not provide much guidance on MTFs within ships apart from the need for one, and for an alternate medical space within a safe area, and high level system requirements in way of MTFs.

AJP-4.10 – Allied Joint Doctrine for Medical Support - Edition B Version 1 + UK national elements [5], is the main NATO publication of interest to ship design. It is the overarching medical doctrine for UK Defence Medical Services and is a hybrid publication; including elements from the UK joint doctrine into the NATO AJP. With respect to MTF design, it provides only high level guidance according to NATO Medical Role (see Table 1 for Role definitions). It also provides a list of core, enhanced and complementary medical modules and stipulates that a Role 2 Basic (R2B) would be expected to contain all the core modules (see Table 2 for list of core modules)
However, no guidance is specifically provided into what modules would be required for a Role 2 Afloat (categorised as a Role 2 Enhanced within the AJP.4.10), nor into the minimum areas or layout of the modules within the facility. This was discussed in depth with RNMS during the workshop and the guidance provided can be viewed in Section 3.3.
Note that whilst AJP-4.10 aims to encourage commonality of equipment, standards and procedure amongst NATO nations, it is more than apparent from the comments of the participating nations within the document signature sheet, that in reality this may be hard to achieve. Each nation may have both differing medical standards and procedures and interpretations of the doctrine.

The implication here is that the guidance for the design of a R2A MTF within this paper may be appropriate for RN vessels but not another nation's navy. Hence designing in flexibility is essential to MTF design for general export as well as for future-proofing.

There is also an extensive framework of NATO Allied Medical publications (AMedPs). The majority of guidance provided within these documents is to do with medical procedure, operations and equipment and of little help to ship design.

3.2 (b) NHS Standards

There is a full suite of NHS guidelines, standards and procedures, which can be used to cover the shortfall in guidance of the existing Maritime standards on medical facilities, in particular for staffing ratios. However, whilst it is left to the host nation to provide the specific guidance for MTFs, it is recognised that the level of care that can be provided on an afloat MTF will not be to the same standard as a national health service (e.g. UK NHS) and operational circumstances may require deviation from laws, rules and requirements set out by national or international organisations.

3.2 (c) Defence Standards:

- DEFSTAN 02-106 Part 1, Part 2 & Part 3 [7,8 &9]

Existing RN Ship MTFs have all been designed to the DEFSTANs [7,8 &9], which provided detailed guidance on the location of MTFs on board, the amount and type of facilities to be provided and the size, overall layout and compartment layout - all according to Vessel Class. It should be noted that Vessel Class is completely different to NATO Medical Role number. Vessel Class is determined by vessel size and type (see Table 3), and dictates the extent and size of Primary Healthcare and Deployed Hospital Care facilities required on a platform. However, the DEFSTANs are out of review date and currently under review. Thus, at present, any naval architect designing medical facilities for future UK RN vessels is driven to use an out of date set of standards, which may be subject to substantial change.

3.2 (d) Commercial Statutory Maritime Standards:

- MGN 482 : Maritime Labour Convention 2006: Medical Care
- MGN 483: Maritime Labour Convention 2006: Crew Accommodation

In addition, the RFA ships (e.g. LSDA) are subject to international commercial statutory requirements (IMO, ILO etc.) as laid down into national law by the Maritime and Coastguard Agency (MCA). These provide limited guidance on ratios of hospital beds to complement and high level system and sanitary requirements for the hospital.

3.2 (e) Casualty and Staff Numbers

No readily available maritime guidance could be found, which provided information into requirements for the number of casualty beds and associated staffing numbers according to Medical Role number. This is required so that adequate medical staff accommodation and ward beds are provided within the ship. To gain an idea of this, NHS standards can be referred to; however application of these standards to afloat facilities may be both onerous and unfeasible. Some guidance into R2A casualty bed numbers and staffing was found within an article in a copy of the Journal of Royal Naval Medical Service [3]. There was, however no source referenced for these figures. This was discussed with RNMS during the workshop.

3.3 (INTERIM) DESIGN GUIDANCE

Prior to the DEFSTANs being released there is a shortfall in available information for the design of MTFs. Hence, the following interim guidance was provided by RNMS :

- A large amount of work has recently been carried out into updating the UK Medical doctrine. Where achievable and within resources, UK Armed Forces will be providing NHS standards of care, but in some circumstances this will be the provision of the best level of care with the available resources and constraints of operating environments.
- Extensive analysis work has been carried out into the need for more medical role afloat facilities. Noted that land based facilities are hard to move vs afloat which are readily deployable and relocated to meet operational requirement.

3.3 (a) Standards for MTF design

- Those specified within section 3.2 of this paper were confirmed as the key standards for MTF design on naval ships.
- DEFSTANs [7,8 &9], noted as out of date and currently under review to be aligned with current UK doctrine [5] and standards– updated versions are due in 2019.

- Examples where the DEFSTAN requirements do not meet national guidelines or standards include standards of operating theatre lighting and use of bunk beds, which should not be specified for sick wards.

3.3 (b) Further SME advice for facilities design

- Clear separation between Role 1 and Role 2 facilities must be provided to avoid conflict of use.
- Bespoke Role 2 IT/ office planning/training space should be provided.
- 360 ° accesses should be stipulated around ITU/HDU beds and Operating and RESUS tables.
- A bath with 360° access (for the treatment of hypothermia) should be provided within the MTF.
- Table 2 shows guidance on NATO modules required for a R2A facility.

3.3 (c) Staffing Ratios & Accommodation

- This information is not open source and thus not readily available to the ship designer unless requested with justification.
  However, typical staff type and number has been provided in table 2 for a R2A MTF. This is similar to that provided in [10] with the addition of another surgeon and an additional ODP.
- Role 1 facilities typically require one medical officer and two medical ratings. However, larger platforms such as CVS may require more.

		Role 2 Afloat (R2A)
MAX Casualty Nos	T1 - Immediate Treatment Group	2
Number of beds	TOTAL	6
	Resus (ED)	2
	ICU	2
	High / Medium Dependency	
	General Ward	2
STAFF	TOTAL	20
	EMERGENCY PHYSICIAN	1
	GENERAL SURGEON	1
	ORTHOPAEDIC	1
	VACULAR SURGEON	1
	SPECIALIST SURGEON	1
	ANAESTHETIC	2
	DENTIST	
	PHYSIOTHERAPIST	
	MITC Medic	1
	MITC Nurse	1
	ED NURSES	2
	THEATRE NURSE	0
	ICU NURSE	3
	Radiographer	1
	ODP	3
	Laboratory Technician	1
	MDSS technicians	1
FACILITIES - Core From NATO Standard AJP-4.10	Emergency Department (Resus)	x 2(T1)
	Initial Surgery Response Capability	x see op theatre
	Specified Diagnostic Capabilities	x
	Patient Holding Area	x
	Post OP (High -medium depndncy)	See ICU
	C4I (inc tele med support)	x
	Medical Supply (Stores)	x
	Medical operations meetign room	x
FACILITIES - Enhancing From NATO Standard AJP-4.10	Imagery	x
	CT-Scan	
	Surgery (Operating	1
	Dental	mobile unit as req
	Sterilisation	x
	Ward (general)	x
	Primary HealthCare (treatment room)	x
	Intensive Care Unit (long term ventilation)	2
	Pharmacy / Dispensary	x
	Laboratory	x
	Mental health	
	Internal medicine	
	Isolation Ward	
	Hospital management	
FACILITIES - Complimentary From NATO Standard AJP-4.10	Additional Clinical	
	Specialist Surgery	
	Oxygen Production	
	Preventive medicine	x
	Hyperbaric Medicine	
	Telemedicine	x
	Transient / Response Ambulances	x
	Magnetic Resonance	
	Frozen Blood Product	
	Frozen Plasma	x
	Animal Care	
	Mortuary	fridge
	CBRN (Decontamination & Treatment)	x
	Physiotherapy	

Table 2. R2A Example staff numbers and NATO modules.

- Demarcation between ranks to be assumed 50/50 split between officers and ratings for first approximation.
- For maximum flexibility, staffing accommodation can be provided in 4 bed bunk cabins. If occupied by officers, only 2 will occupy the cabin, if by ratings 4 will occupy. RFA cabins to be ensuite.
- Deployment of increased tele medicine has changed the types of people deployed, e.g. the radiologist may no longer need to be on board the platform as imagery can be sent to them for analysis. Increases in augmented reality technologies may also change the way operations are conducted and the number and skill sets of staff required. Facility requirements for tele medicine can be accommodated within existing spaces as this technology only requires integrated CIS and appropriate hardware. It should be noted that sufficient bandwidth within the ships transmission system to cope with large medical files and data streaming would also be required.

4. EXISTING VESSEL COMPARISON

To further understand and contextualise the current R2A arrangements within the UK fleet, Table 3 shows the list of current and planned RN/RFA vessels and their designated medical facility role. It shows the spread of platform to which R2A facilities are currently fitted and also highlights that the newer platforms such as the T26 are likely to have a higher medical capability compared to their predecessors.

The current in service vessels which have R2A facilities are all getting close to the end of their service lives and are likely going to require replacement over the next 10 – 15 years.

It can therefore be concluded that if the number of R2A facilities is maintained at the same level within the fleet the replacements for classes such as LSDA, AOR, and LPH will require R2A facilities.

As stated within the previous section, the Defence Standards are out dated and no longer represent the best guidance for medical facility design. To highlight the disparity Table 4 shows a comparison between the out dated DEFSTANs, NATO designated spaces and the current vessels upon which have R2A facilities.

In addition, a discussion with Medical SMEs has provided feedback on the good and bad points associated with the facilities identified above in Table 4, from which the following main points were distilled:

LPD: the facilities on HMS Albion and Bulwark are a good example of a role 1 facility. However access to the facility from the fight deck and vehicle deck is via single width watertight doors and requires a considerable amount of effort with a casualty on a stretcher.

LPH: The facilities on board HMS Ocean provide a good arrangement and the operating theatre size is adequate. However there are conflicts between the Role 1 and Role 2 Facilities, i.e. if the medical spaces are being used in

Role 2 then there is restricted access for Role 1. There are also no facilities for office /training /planning and IT.

LSDA: The issues with the LSDA are similar to that of the LPH arrangement, with conflicts between the Role 1 and 2 facilities on board. The LSDA does however have very good access to the medical facility from the fight deck.

A further piece of work is to add the QEC design areas to the comparison however these were not available at the time of the study.

Vessel	DEFSTAN Class	Medical Role
Aircraft Carriers (QE Class) (RN)	A	2
Landing Platform Helicopter (RN)	B	2
Landing Platform Dock (RN)	B	1
Type 45 Destroyer (RN)	C	1
Type 23 Frigate (RN)	C	1
Type 26 Future Frigate (RN)	C	2
Type 31e Future Light Frigate (RN)	C	1
Mine Countermeasures Vessels (RN)	C	1
River Class Patrol Vessels (RN)	D	1
Fast Inshore Patrol vessels (RN)		0
Auxiliary oiler and Replenishment Fort Class (RFA)		2
Fast Fleet Tankers (Wave Class) (RFA)		1
Forward Repair Ship (RFA)		1
New Tankers (Tide Class) (RFA)	No DEFSTAN CLASS Identified	1
Tanker (Wave Class) (RFA)		1
Landing Ship Dock (Bay Class) (RFA)		2
Primary Casualty Receiving Ship (RFA)		3
Solid Support Ship (Fort Class) (RFA)		1
Future Solid Support (RFA)		2 (expected)

Table 3: List of current & Future UK Naval Fleet and their respective levels of Medical Capability and Vessel Class

5 ROLE 2A BASELINE FACILITIES

The list of modules, facilities and staff and casualty bed numbers, which should be provided for a R2A vessel can be seen in Table 2. However, the lay out of an indicative R2A baseline MTF allows for a clear illustration of how this guidance can be best applied to the design of MTFs. A discussion with RNMS during our workshop provided the arrangement shown in Figure 1. This arrangement has been notated with the area allocation for each compartment.

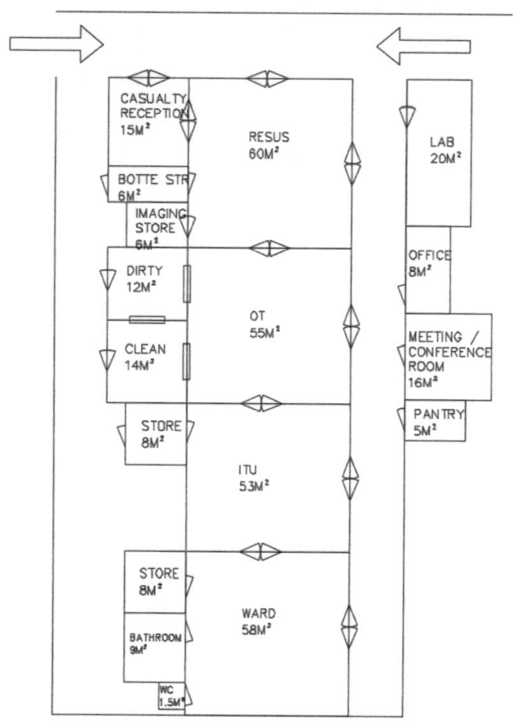

Figure 1: Role 2 Baseline Facility

The Space allocations for each of the areas within Figure 3 have been predominantly derived from the NHS standards [12 & 16], bearing in mind the 2-1-2 bed requirement of the RESUS, OT and ICU respectively. In discussions during the workshop with RNMS it was clear that the Defence Standards were inadequate in their space provisions and when a comparison was drawn between in service vessels, Defence Standards and the NHS Standards there were many inconsistencies.

The NHS Standards provide a good target as they account for a variety of equipment and enable a degree of flexibility within the space envelopes. There are however, some discrepancies for example the ward bathroom size, ready use stores and laboratories.

- The bathrooms assigned within the baseline is larger than the NHS guidelines as it is assumed that it will provide both a shower and bath with 360° access.
- The ready use stores are larger than the NHS provision as reference [11] stated only 1m^2, it was assumed that as land based hospitals have the ability to locate cupboards within other spaces adjacent to the wards, these are used as stores, therefore a larger dedicated ready use store within the ward is not necessary.
- NHS hospital facilities have an extensive set of laboratories, the HBN guidance for laboratories [12] quotes over 3000m^2 for all labs and c220m^2 for the Microbiology and Blood Labs. For the purposes of this study it has been assumed that a single Pathology lab of 20m^2

would be sufficient based on figures for the PCRS (RFA Argus).

It should be noted that Figure 1 shows dedicated Role 2 facilities only. As MTFs should also incorporate the capabilities of lower roles, a Role 1 facility would also be required within the MTF design, in addition to and separate to the Role 2 facilities.

6. CASE STUDIES: REVIEW OF MEDICAL FACILITIES ON TWO MRV CONCEPT DESIGNS

Two case studies of MTFs on multi-role vessels of different size and medical role have been developed and were discussed with RNMS in order to derive further guidance on MTF design and show how it can be applied.

6.1 120M MULTI-ROLE VESSEL

The first vessel consists of a 120m Multi Role Vessel. The primary role of the vessel is patrol of EEZ waters, however it also incorporates a helideck for a medium weight naval helicopter, austere accommodation for up to 100 EMF and a large hangar / garage on the same level as the flight deck, which can be used to store containers, transport military vehicles or used as a mission bay with modular containerised units.

Length: 121m
Beam: 19m
Displacement: 4700te
Speed: 24 knots
Crew: 60
Austere Accommodation: 100

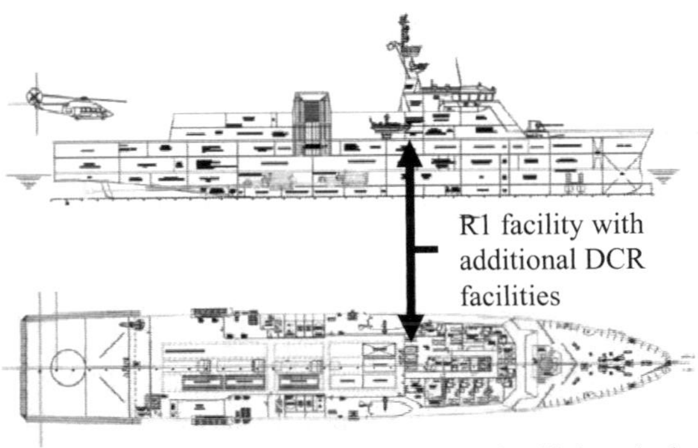

Figure 2: 120m MRV profile and flight deck arrangement.

The original MTF comprised a standard commercial ship sick bay and consulting office (Role 1), designed for export rather than a RN vessel, although some guidance had been taken from the (currently under review) DEFSTANs in designing the arrangements. However, further information received on the ship's requirements implied the need for some Deployed Hospital Care (forward) capability to support peacekeeping operations. As an illustration of the issues faced when integrating MTFs into multi- role vessels: the space on this deck is also required for the helideck, extended garage/hangar, aviation facilities, vehicle and dive workshops, boat bays, boarding gear, boat crew and diver change changing facilities, gun and mooring deck. All these form essential parts of the primary role of the ship and as such, accommodating a full R2A facility proved difficult. Instead, a large 3-4 bed RESUS bay, to provide casualty reception, enhanced fieldcare and DCR prior to MEDEVAC, was incorporated with a direct route to the flight deck through the extended hangar.

These facilities were discussed with SMEs and the key points are summarised below:

- Specific medical laundry is not required unless the size of the MTF takes the ship's laundry over capacity. (Clinical laundry is placed in special bags with the rest of ship's laundry).
- A medical office / planning room is essential for all naval ships
- A controlled temperature medical store must be provided in addition to ambient temperature stores.
- Laboratory space is required.
- Stairs accessing other areas of the ship (e.g. accommodation) should not open straight into medical areas. (Clear separation of medical facilities from core ship function is required to ensure cleanliness standards are maintained within the MTF.)
- 360° access is required around the bath.
- Medical gas storage should not be on an external boundary to avoid extremes of temperature.
- The RESUS space should support fewer beds in order to provide adequate access around beds for the required number of staff.

The revised 120m MRV MTF arrangement can be seen in Figure 5. This MTF is still only R1, as it has no surgical capability. However, the MTF could be revised by replacing the large resuscitation bay with a 1 RESUS bed/ 1 OT / 1 ICU bed DCR/DCS facility. Whether this could be considered a R2 Light facility would need to be reviewed. (During discussions on existing vessels it was established that the T45 could more or less accommodate an arrangement such as this).

6.2 180M MULTI-ROLE VESSEL

The second case study investigated was a 180m Multi-Role vessel. This concept design is for a generic multi-role vessel which can facilitate modern navies in undertaking multiple mission types using a single re-roleable platform, with interchangeable modules. The design is being developed within the Babcock Concept Design Team and has the following principal particulars;

Length: 180m
Beam: 32m
Displacement: 26,000te
Speed: 25 knots
Crew: 90
Accommodation: 400

The key features of this concept are the large flight deck, LCU stern dock and approximately 2500m^2 of mission bay. The design has vehicle access to the dockside via a side boarding door and ramps between each of the vehicle decks and up to the flight deck.

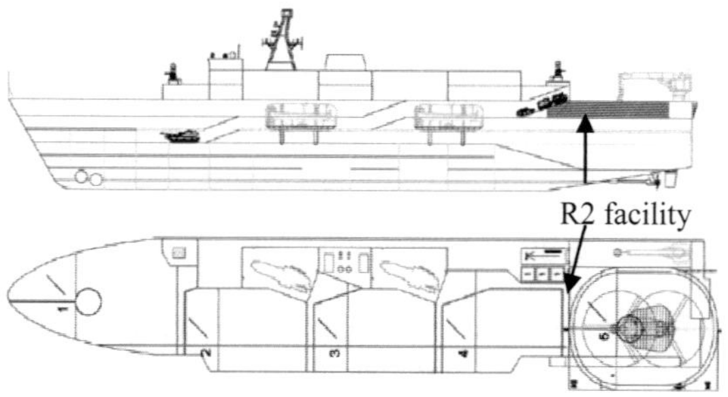

Figure 3: 180m Multi Role Vessel

Figure 3 shows a simplified profile and upper deck arrangement for the 180m MRV.

The space allocated to the medical facility is situated on 2 deck and annotated in Figure3. The intention of this design was to have a permanent medical facility which could then be expanded with the use of modules within the confines of the mission bay spaces directly aft of the facility. Two versions of the MTF were initially produced for discussion, one based on the DEFSTAN [7&8] and one on the LSDA R2A facilities. However, it was decided that it would be more constructive to receive guidance on an ideal R2A baseline facility (see section 5), rather than review further pros and cons on each.
As this is a new concept design, the medical facilities can be thought about early on and adequate space provision can be captured. This allows for the R2 baseline facility to be fitted to the ship and orientated in a way which allows easy access to the reconfigurable mission space.

The case study design was discussed with RNMS and the principal issue identified was access to the medical facility via ramps. The design provides ramps between the flight deck and mission bay where the medial facility is located. This solution however would not comply on the grounds of health and safety due to the hazards associated with transporting stretchers up and down ramps whilst the ship is moving. Instead discussion was held over the ability to move the medical facility closer to the primary point of casualty embarkation. However, this would lead to compromising the facilities ability to expand in capacity using the space with the mission bay. The design also has three means of allowing casualty access each of which are on different decks, therefore wherever the facility is sited would compromise the other access points. It was therefore concluded that the use of lifts between the decks would be the most appropriate means of transporting casualties from the point of embarkation to the medical facility.

7. MODULAR TECHNOLOGIES

7.1 GENERAL

The continued development of modular medical technology means that the ability to successfully treat injury and illness away from permanent conventional land based medical facilities is constantly improving.
NATO's plan to mitigate shortfalls in medical facilities includes a 'Modular Approach – ModA' with the aim to pool standardised medical modules amongst NATO nations for supply where and as required [5].
Technologies and techniques developed on the battle field and during recent conflicts such as Afghanistan are now being widely used across the military. Notable examples are the modular, deployable hospital facilities such as the Marshalls Deployable Field Hospital, trialled by the MOD in 2015[13], which provides a capability for NATO to deploy a CBRN Protected Role 3 hospital anywhere in the world at relatively short notice.

Whilst tented hospitals have their place, there are many advantages in the use of containerised 'hard' modules over a tented hospital. The following are just a few:

- Modules comply with ISO standards for intermodal transportation and so are readily transported by road. They are compatible with existing container handling devices found in rail and ship transportation;
- Complex medical and other equipment can be permanently integrated within the module reducing setup burden and into action time;
- Optimal internal temperature and humidity can be readily maintained irrespective of external conditions. Ingress of dust and other contaminates can be minimised providing enhanced cleanliness for clinical areas

Containerised modular units such as scientific laboratories, refrigerated stores, accommodation blocks and ROV control rooms have already been used successfully on board commercial ships, in the oil and gas and research sectors, for several years.

Another key part of this investigation is to consider how these containerised modular medical facilities could be used on board naval ships (e.g. within flexible mission spaces, which are increasingly being designed into naval ships) to enhance R2A facilities and tailor the facilities according to mission type (see Section 8).

Examples of these modules are shown in Figure 4.
It is worth noting that these modular medical units are not designed for use on a maritime platform and there are a number of issues to be addressed, not least meeting

naval ship fire standards and gaining recognised maritime certification and Authority approval prior to use. However, none of this is unsurmountable with sufficient lead time and given that the modules can be designed to meet specific requirements.

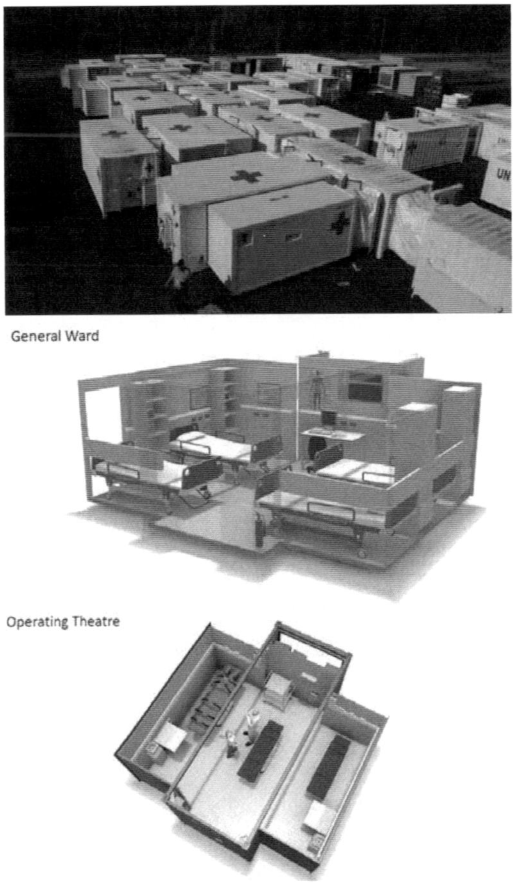

Figure 4: Marshall Aerospace and Defence Group Containerised Field Hospital and Modular Units

7.2 UK RNMS FEEDBACK

Extensive research into the use of modular medical facilities on UK naval platforms has been carried out by RNMS. The following should be noted:

- Augmenting a R2A facility up to R3A standard using modular facilities would be unfeasible due to the sheer volume of containers that would need to be accommodated, and stored. (A land-based R3 field hospital involves over 100 containers)
- However, augmenting a R2A using a few simple medical or accommodation modular units to 'enhance' the MTF for particular missions in terms of capacity (e.g. providing support to Littoral Missions), could be beneficial.
- They could also be used to provide basic mission specific and segregated medical facilities for HADRO or NEO.
- The units used would ideally not require any special storage requirements.
- Modular units will require manufacture and testing to maritime standards in order to be utilised within the ship. They will also need acceptance from the appropriate Maritime Authority.
- Due to shortages in man power and a time to effect of 48 hours, the forward basing of modules may be appropriate. It is recognised there are a number of logistical issues that would need solving to make the use of modules appropriate.

8. MODULAR CASE STUDIES

8.1 VIGNETTES

To help provide context to the use of the medical facilities within a multi-role platform, two vignettes were created;

- HADRO/NEO (Mass Casualty Evacuation)
- Littoral Operations

It should be noted that there are a number of suitable vignettes which could be explored for these types of platforms and a vast number of permutations between them. The exercise of producing a limited number of vignettes was to instil debate during the workshop and further understand the likely scenarios which may be encountered and where the limitations of the R2A medical facility would lie.

Following discussions with the SMEs, a significant point was made:

It is against Navy Medical doctrine to bring civilians into the ships medical facilities. The reasons for this are due to an obligation to provide a continued duty of care to military casualties. This and the low capacity nature of a maritime R2A baseline MTF means that only a small number of civilians would take the facilities to capacity, making it redundant for military use. Therefore just expanding the Role 2 facilities, with e.g. more ward space, would not be useful in a HADRO scenario as facilities for civilians would have to be standalone and completely separate from the ships MTFs.

However, there were a number of options discussed where supplementing a R2A MTF with modular facilities would be appropriate and the use of modules to provide a standalone facility for HADRO was also discussed.

8.2 LITTORAL OPERATIONS

As the medical facilities would be used for military personnel, the enhancement of a R2A facility would be appropriate under a Lit'M scenario. The key enhancements that could be required are, additional:

- ward space,
- medical stores,
- personnel accommodation,

- sanitary facilities,
- medical waste storage,
- rehabilitation facility, and
- additional medical operations and planning facilities.

Having the ability to add these facilities would greatly enhance the capacity of the R2A facility.

8.2 (a) 180m MRV Modular Arrangement for LIT'M

Placing the MTF on the same deck as the reconfigurable space allows for easy augmentation with the use of modular systems. Figure 6, illustrates how the facility can be augmented if it were required to undertake a Lit'M operation.

8.3 HADRO/NEO

HADRO/NEO was discussed under the context of needing to embark a large number of civilian personnel and transit with them for a short period. As previously discussed these civilians would not be allowed access to the on board medical facilities, therefore a separate facility would be required within a mission bay. Being in close proximity to the ships medical facilities would however be beneficial from a planning and organisational context.

The key facilities required would be;

- Reception and triage
- GP consultation rooms
- Minor ops room
- Cot bed spaces
- Sanitation facilities
- Seating / waiting area
- Stores medical
- Stores victualling and general aid supplies
- Morgue

A number of other considerations would also be required. For example: the need to contain the civilians in a single area and prevent access to the rest of the vessel; the ability to charge a large number of mobile phones and provision of civilian WiFi; and a clear barrier would also be required between the ships medical facilities and the civilians embarked.

8.3 (a) 120m MRV Modular Arrangement for HADRO

The MRV 120m has a large, flexible hangar/ garage/ container store located just aft of the MTF and leading straight onto the helideck. This is an ideal space to be used for modular facilities. The deck will be constructed with a relatively dense grid of container securing twist locks to allow some flexibility in the arrangement of modules (or stowage of containers). Figure 7 illustrates how medical modules have been arranged for HADRO within the garage.

Notes:

- There should be only one point of entry to the facilities for civilian casualties.
- The planning room with reception desk to process documentation required is positioned at the entrance to hangar.
- The holding area is forward of this within the hangar (portable stretcher beds, seats etc.).
- A barrier is located forward of the holding area and aft of this general GP surgery for civilians, minor operations room and general ward is positioned.
- This is followed by a second barrier, medical stores and a morgue.

8.3 (b) 180m MRV Modular Arrangement for HADRO

There are a number of possible arrangements of the HADRO facilities within the context of the 180m MRV, Figure 8, shows one such arrangement. This arrangement keeps the civilians separate from the on board MTF using a barrier and provides a self-contained on-board space for their carriage to a safe port.

The facilities include;

- Triage and general GP surgery for civilians.
- Minor ops room providing essential minor surgery including delivering babies.
- Small patient holding area for civilians requiring medical supervision.
- Mortuary and stores located separate but within close proximity to the holding facility.

9. CONCLUSIONS

It is clear from the work undertaken that the R2A platform is of extreme strategic importance and key to casualties receiving lifesaving surgery and treatment within the appropriate timescales. Our work has shown how there is likely to be a decline in the current UK R2A facilities as vessels leave UK service over the next few years (and noting LPH HMS Ocean has already been sold).

It is evident that clear standards and guidance need to be available to allow the engineering community to provide facilities within their new designs that meet the end users requirements. The provision of interim guidance from medical SMEs and an indicative R2A baseline design within this paper will help the ship design community prior to the official release of the DEFSTAN in 2019. It is hoped that this information will help promote best practice, such as clear separation of R1 and R2 facilities within the MTF, flexibility and 360° access around R2 beds, into MTF design. It is also key that any new medical facility design is incorporated at the earliest stages of a platform design. Dialogue between the designers, authority and end users also needs to be established to ensure these facilities are capable of meeting future needs and mitigate obsolescence during a platform's lifetime.

In addition, the 'flexible mission space', present in many modern and future naval ship designs, and the advance in modular technology have the potential to benefit the capability of the platform MTF. If the UK continues to use its naval platforms to contribute to humanitarian aid and disaster relief operations, then the use of simple medical containerised modules within mission bays in future platform designs will further enhance the RNs capacity and capability to help without compromising the integral MTFs.
It is acknowledged that this paper presents only an initial investigation into the possibilities of medical capability enhancement through modular facilities. Further work to consider the associated cost impacts and logistics is required in order to gain the full benefit of these solutions.

10. ACKNOWLEDGEMENTS

The authors would like to acknowledge the help provided by: Lucy Lowe (Babcock Graduate) and John Abbott from Marshalls Aerospace and Defence Group for their help at the workshop; Commander Kay Hallsworth and the Medical Operational Capability organisation within NCHQ , for their support of this paper and willingness to accommodate the design workshop. We would also like to thank a number of individuals within MOD DE&S CSS and In-service support in providing permission to use existing ship arrangements for comparison.

11. REFERENCES

1. "The rise of the multirole vessel (MRV)", Gareth Evans, Naval Technology 24 January 2012 (https://www.navaltechnology.com/features/featurethe-rise-of-the-multi-role-vessel/)
2. House of Commons Debate Pack - ' UK's Contribution to International Disaster Relief' - CDP-0030, 2 February 2018
3. ' Emergency Medicine in Role 2 Afloat', S Doherty, S Gilbert, D C W Henning, Journal Royal Naval Medical Service 2013 Vol 99.3
4. SO1 Medical Doctrine DCDC, 'The Operational Patient Care Pathway', JSP 950 Part 1, CH4, Lft 1-4-1 V2.1, November 2014
5. AJP-4.10 – Allied Joint Doctrine for Medical Support - Edition B Version 1 + UK national elements MAY 2015
6. ANEP 77 – Naval Ship Code
7. DEFSTAN 02-106 Part 1 – Common Requirements for Medical & Dental Organisation in HM Surface Ships and Submarines - UNDER REVIEW
8. DEFSTAN 02-106 Part 2 – Specific Requirements for Medical & Dental Organisation in HM Surface Ships and Submarines UNDER REVIEW
9. DEFSTAN 02-106 Part 3 - Requirements for Medical Organisation for Action in HM Surface Ships UNDER REVIEW
10. Special feature - International mine countermeasure exercises 2013: Role 2 Afloat - JRNMS 2013 Vol 99.3
11. NHS Estates, "Facilities for Surgical Procedures: HBN 26, Volume 1, NHS Estates Best Practice Guidance
12. Department of Health, "Facilities for Pathology Services", Health Building Note HBN 15, 23rd April 2005
13. Marshall Aerospace and Defence Group, 'Best In Class Medical Treatment Facilities', Brochure, www.marshalladg.com, 27th December 2018.
14. MGN 482 : Maritime Labour Convention 2006: Medical Care
15. MGN 483: Maritime Labour Convention 2006: Crew Accommodation
16. Department of Health, "Critical Care Units", Health Building Note HBN 04-02, 2013

12. AUTHORS BIOGRAPHY

Pilar Morales is a Principal Naval Architect within the Naval Ship Concept Design Team in Babcock. She is responsible for leading and contributing to a range of Babcock naval ship concept designs and bids. She has worked for Babcock for 17 years in design roles supporting refit and build of naval surface ships, submarines, super yachts and commercial vessels.

Harry Schweidler is a Senior Naval Architect working as part of the Babcock Concept Design Team in Bristol. He has been at Babcock for 10 years and has gained experience working in a large variety of roles from concept through to disposal. Most recently he spent four years seconded into the MOD NDP before moving to the Babcock Concept Team at the beginning of 2018.

	Compartment	NATO Role 2 B Spaces	Defence Standard: Class A (CVS) Area (m2)	Defence Standard: Class B (LPD/LPH) Area (m2)	In Service RN Vessels: LPD (Class B) Role 1 Area (m2)	In Service RN Vessels: LPH (Class B) Role 2 Area (m2)	In Service RN Vessels: LSDA (Class B) Role 2 Area (m2)	In Service RN Vessels: Fort Victoria (No class) Role 2
Defence Standard	**Lobby / Waiting area**	x	21	16				
	Surgery	x						
	Treatment room / Dispensary		12	12	28	10	20	
	Operating theatre		17			23	25	17
	Sterilising room		4.5	4.5		7		
	Main Ward	x	46	28	27	58	43	39
	Ward Wash Room		3	3				
	Ward Bathroom		4	4	5	9	10	3
	Ward WC		1.5	1.5	2	6	2	2
	Ward pantry					5		
	Isolation Ward / Special Ward		10			17	23	28
	Reception		7.5	6.5	6	12		
	Medical Store	x	18	18	17	64		8
	Consultation room		8	8	7	16		9
	Dental Surgery		16	16	15	20		
	Dental Lab		10			9		
	Pathological Lab / Dark Room	x	10					
Other	Casualty reception					6		13
	Resuscitation					15		
	Ward Sluice					7		
	Special ward shr/WC					6	3	
	General use WC				2			
	Medical Gas bottle store				6			
	Dental waiting room					8		
	Misc Store							
	Total Area		**306**	**160**	**115**	**256**	**126**	**119**

Table 4. Medical Facility areas comparison

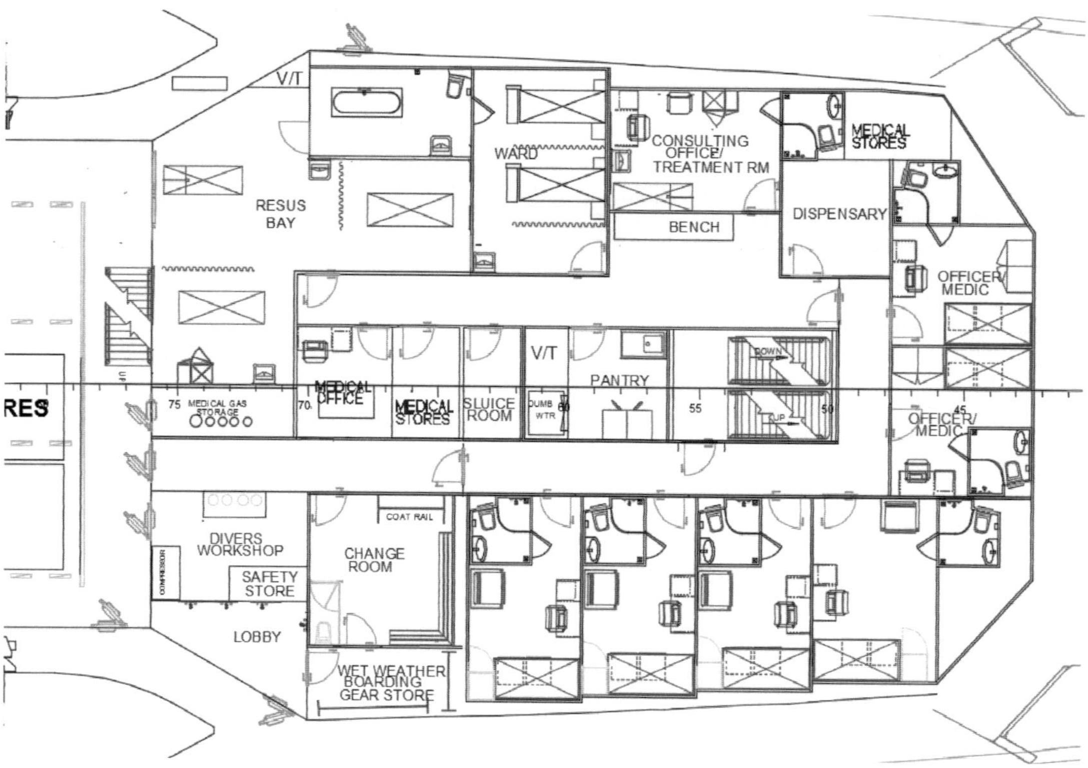

Figure 5: 120m MRV Medical Facility

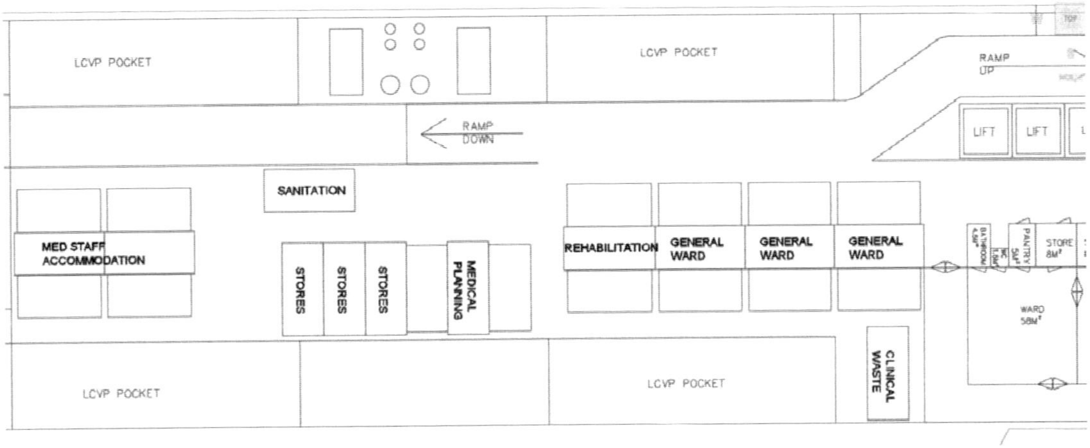

Figure 6: 180m MRV Modular Medical Arrangement for Littoral Operations Support

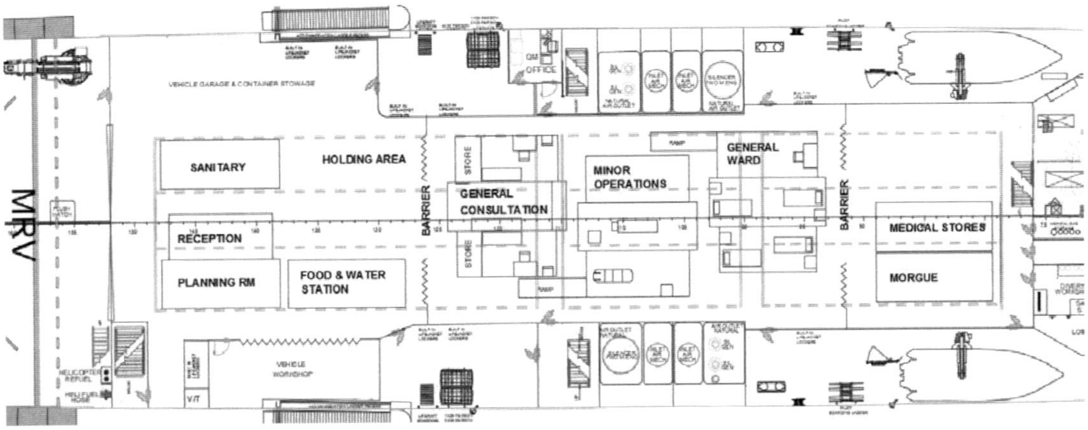

Figure 7: 120m MRV Modular Medical Arrangement for HADRO

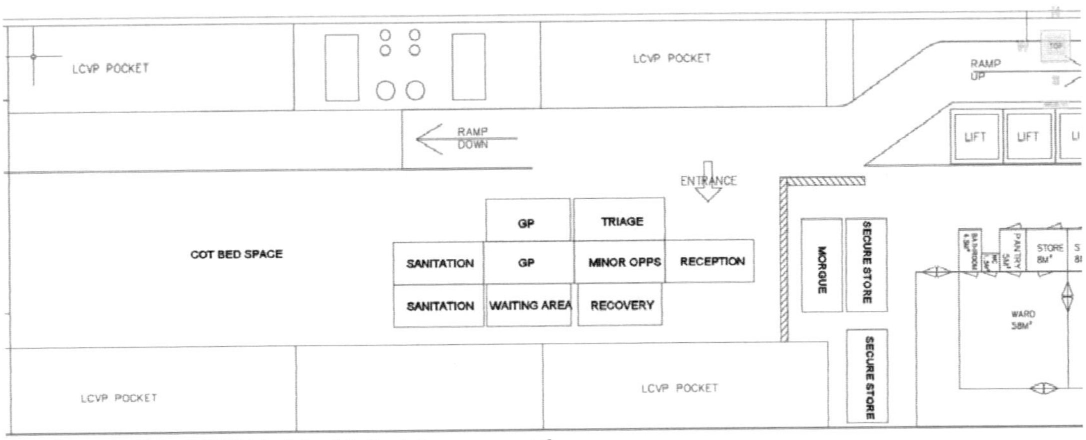

Figure 8: 180m MRV Modular Medical Arrangement for HADRO

THE FUTURE WARSHIP – GREEN AND MISSION CAPABLE

C Lyster and **A James,** Babcock International, UK

SUMMARY

The environment and its protection are uppermost in the public consciousness as we progress into the 21st Century. There is an increasing need for modern navies to consider environmental protection. This protection starts on the drawing board, at the very beginning of warship development.

This paper considers warship development and the opportunity to intelligently integrate environmental protection into the design, build, operation, maintenance and disposal of the modern warship.

The penalties for not taking environmental protection seriously are not just subsequent physical damage, but are based in international legislation and reputational damage.

The accelerating pace of change, both in the earth's climate and in legislation, is going to be an ongoing issue for warship designers. This paper introduces the challenge and opportunities associated with environmental protection. The need to focus on today's legislation and anticipate the requirements of tomorrow are at the forefront in developing through-life ship solutions.

1. INTRODUCTION

Modern warships have a great many demands placed upon them. They must have the ability to face asymmetrical threats (air, surface and underwater) which requires a variety of weapons, computers and sensors. In addition they must be able to carry out rescues and support humanitarian missions while representing their country around the world. Apart from all of these disparate capability requirements they must also be "green." Being "green" requires the ship to have a minimum impact on the environment firstly by not discharging pollutants into the environment, secondly to have a low energy signature and thirdly to be easy to recycle at end of life. This paper explores current requirements for being "green" while considering future requirements (for the immediate future, through life and end of life).

2. WHY BE GREEN?

In the Introduction it was mentioned that a warship has a special requirement of representing its country around the world. As a result a warship cannot be restricted in its operation by not complying with environmental regulations. In the past warships had a certain "immunity" from pollution regulations. As an example the Canada Shipping Act states that "Government vessels not engaged in commercial operations are exempt from MARPOL" (Ref 1). However local regulations (including local port bylaws) may take precedence and as a result a warship may be deemed to be non-compliant. This could result in a warship being refused entry for provisioning or recreation after a long stressful mission. It could also be expected that, in the foreseeable future, stricter environmental controls and considerations may be imposed by IMO, MOD or other marine environment stakeholders.

3. REGULATORY REQUIREMENTS

NATO warships in general sign up to the Naval Ship Code (NSC) which includes the Allied Naval Engineering Publication (ANEP-77) (Ref 2). Although the NSC is primarily a goal based set of regulations which achieves safety and environmental compliance through the practice of As Low As Reasonably Practicable (ALARP) it has certain references to outside prescriptive regulations. As an example under ANEP-77 Performance Requirements (6.2) it states "Where possible the IMO Guidelines on Ship Recycling are to be followed, as provided in IMO Resolution A.962(23), 5th December 2003." Note that ANEP-77 does not oblige members to follow the IMO Guidelines at this time.

4 ASPECTS OF ENVIRONMENTAL PROTECTION

There are 5 main aspects related to ship pollution which are as follows:

1. Emissions into the air (NOx, CO_2, heat energy, SOx)
2. Pollution of the ocean (petroleum products, sewage, ballast water, paints)
3. Noise, vibration, radiation (airborne and underwater)
4. Energy Loss
5. Hazardous Materials

The sources of each of these pollutants are shown in Figure 1. They will be discussed in the following section.

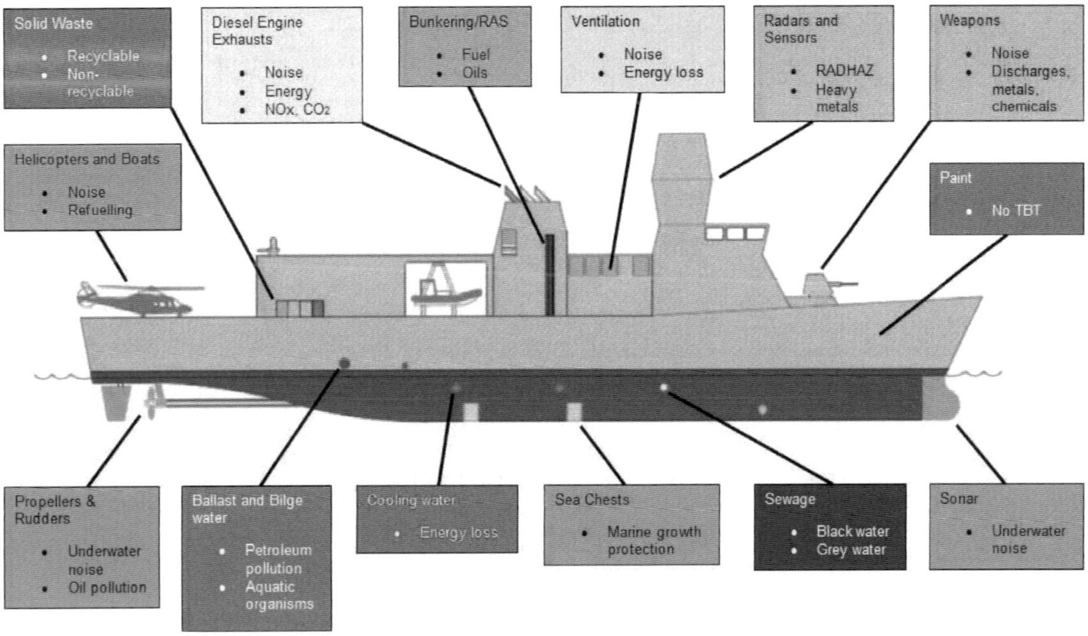

Figure 1 – Warship Pollutant Sources

4.1 AIR EMISSIONS

As listed above there are 4 areas of emissions that are considered, nitrogen oxides (NOx), carbon dioxide (CO_2), heat energy and sulphur (SOx).
Air emissions can be improved using the following methods:

a. Treating the exhaust gas to reduce NOx (such as Selective Catalytic Reduction (SCR))
b. Minimising exhaust gas production (CO_2), by minimising power requirements (which will also minimise the heat energy consumption of the ship)
c. Reducing heat energy losses by having better ventilation systems
d. Sulphur is contained in heavy or residual fuels which are not used on most warships but it is worth noting that legislation is coming into force which obliges fuel suppliers to reduce the sulphur content of these fuels (Ref 3).

4.2 WATER POLLUTANTS

The main groups that are of concern for polluting the water are as follows:

a. Petroleum products (in bilge water, from bunkering operations (spills) and from controllable pitch propellers)
b. Sewage (black and grey water)
c. Ballast water (transfer of non-indigenous species of marine life)
d. Paints
e. Sea-chest marine growth prevention systems (they discharge copper into the water to prevent marine larvae from attaching and growing inside) (Cathelco Ref7)

4.2 NOISE, VIBRATION, RADIATION

There are a variety of sources of the above on a ship and they can be airborne or subsea:

a. Engines, propellers and associated propulsion equipment
b. Weapons
c. Radars, Sonars and other Combat Systems

Engines and propellers can induce vibration and noise into the surface and underwater environments which can affect marine life. Airborne noise can affect personnel living near a port which can prevent the ship operating certain equipment.

Weapons can produce loud noises which can have a very serious effect on marine life as well as being harmful to human hearing.

Radars, sonars, antennas and other sensors can produce a variety of radiation and noises that can affect human and marine life in a variety of negative ways including: causing cancer and reducing fertility.

It should be noted that some of the above sources are intermittent (firing of weapons) whereas others such as ventilation, sonars and radars are operated almost continuously.

4.3 SOLID WASTE

A ship generates a variety of solid waste both recyclable and non-recyclable and this must be stored on board before it can be taken to a suitable shore facility.

- Recyclable waste – Cardboard, plastic, glass, food waste
- Non-recyclable waste – Medical waste, cleaning products, plastic (non-recyclable types), weapons discharges

Recyclable waste primarily covers packaging for food and other products.

Galley food waste is a bit unusual as it fits into both recyclable and non-recyclable categories. If the vessel is over 12 miles from land food waste may be dumped overboard. However if the vessel is closer than 12 miles the food waste must be taken to a suitable shore facility for disposal.

Weapons have a variety of discharges when used. These include:

a. Spent cartridges (brass and other metals)
b. Chemicals
c. Burned residue
d. Plastics
e. Metal residues from projectiles
f. Noise (covered above)

4.4 ENERGY LOSSES

In Figure 1 above the category of "Energy Loss" is highlighted under diesel engines, cooling water and ventilation. In the past warships have not had to consider energy loss in their designs other than when evaluating the range of the ship at different speeds although propulsion systems are now being selected with consideration to reducing fuel consumption by optimising the plant to the vessel's mission profile (Ref 5). However reducing fuel consumption has been driven more by operational budget constraints than for environmental considerations.

Many commercial vessels have used energy recovery systems for years. Some of the systems include:

a. Exhaust gas boilers
b. Turbo-generators (in association with the exhaust gas boilers)
c. Fresh water generators that use engine cooling water to make drinking water

The above systems have not been considered on warships mainly due to volume and weight concerns particularly with exhaust gas boilers as they are high up in the ship. Even vessels that are not front line do not use heat recovery systems. The Royal Fleet Auxiliary (RFA) is a case in point; even the newest class of ships in their fleet, the Tide Class, do not have any heat recovery systems such as uptake boilers or fresh water generators.

A final point is insulation. New regulations may oblige new ships to minimise heat loss and reduce energy requirements for air conditioning by improving the insulation. The easiest way of doing this is to increase the thickness of the insulation. However this will add weight to the ship.

4.5 INVENTORY OF HAZARDOUS MATERIALS

The Hong Kong International Convention for the Safe and Environmentally Sound Recycling of Ships, 2009 states: "Ships to be sent for recycling will be required to carry an inventory of hazardous materials, which will be specific to each ship" (Ref 6).

The IHM was originally introduced to protect workers dismantling ships at end of life. This has evolved and by 2020 all large ships will be required to carry an IHM, not only to protects those dismantling but also those who may be exposed to hazardous materials on board and it allows port authorities to know what is (potentially) on-board any vessels coming in to port. Indeed this could extend further with nations requesting to see the IHM before permitting access to territorial waters and potentially denying access if the vessel is non-compliant with local regulations.

The IHM is maintained throughout the life of the ship. It ensures that when regulations change the equipment with hazardous materials in it can be removed or managed to ensure that environmental safety is maintained to protect personnel and the environment.

It is important to note that some hazardous materials are fixed or built in to equipment while others are consumable. Some weapons and combat systems use heavy metals in them (such as cadmium) and will only become a concern when the equipment is changed or the vessel is scrapped at end of life. Fuels and lubricating oils have different hazards that must be managed in a different way including processes to ensure safety during bunkering, minimising leaks on board and maintenance of tanks to prevent marine growth and corrosion.

5. TRENDS IN ENVIRONMENTAL LEGISLATION AND REGULATION

The major trends are as follows:

a. Changes to Inventory of hazardous materials;
b. Requirements for a higher level of sustainability;
c. Higher public awareness of issues such as plastic pollution;
d. Ship owners remaining responsible even after handing over to ship breakers;
e. Changes in the use of munitions and targets.

5.1 SUSTAINABILITY AND RECYCLING

The Environment can be a contentious subject ranging from some believing at any negative impact on the environment is unacceptable while the other end of the spectrum believe that human activities have no influence on the environment.

So where does the navy fit in? Should we dock all ships shut down all engines and remove all pollutants? Or maybe we should assume that the environment is so massive any single ship couldn't possibly impact the environment and shouldn't even be a considered for a warship?

The truth is that neither of these are correct, the MOD acknowledges its activities impact the environment but equally those activities are the same ones that protect our borders and interests, which provide humanitarian aid and maintain a global presence. The aim is to manage the environmental impact of operations and procurement to a level that is sustainable, but what is sustainable?

"Sustainability is the ability to continue to operate at a particular level for a period of time" (Ref 7) (see Figure 2) quite simply, sustainability is protecting our future interests, it is looking to reduce, reuse and recycle for long to ensure we have the resources for future procurement and operations; of course this has to be approached pragmatically, oil is a non-renewable resource but at this time there is no practical way the Navy can remove the use of oil from its ships. What the Navy can, and is doing is make more efficient use of these resources.

5.2 MATERIALS

The sustainability of a ship is influenced not only in how it is operated but what it is made of, this applies both in service for refits and at the disposal stage.

For example in Bangladesh the steel taken from ships provides the necessary scrap metal for products such as rebar and typically represent a much smaller carbon foot print than smelting the products from iron ore according to S.M. Mizanur Rahman et al (Ref 8).

If we follow the waste hierarchy (Reduce - Reuse - Recycle - Recover - Landfill) in Figure 2, we can see that recycling is actually quite low on the hierarchy, can we reduce what we use?

Reduce - Warships being expensive usually don't have a great deal of superfluous steel but by being intelligent in designing ships to certain operating profiles we can potentially have smaller more optimised ships. If a warship were designed from the outset to have a hull life much longer than its in-service life with the ability to undergo major refit in relative ease this could represent a much more attractive option for reuse or resale and the potential to be more economically viable in the long run.

Reuse - It can be attractive to a navy to reuse a warship by repurposing it to fulfil a different role, or by selling the ship to a foreign navy.

Recycle & Recover - Dismantling ships at the end of their useful life can carry many risks to the environment and safety of personnel. Substances such as asbestos, glass fibres and oil residues can only be managed effectively if their location and quantities are known beforehand.

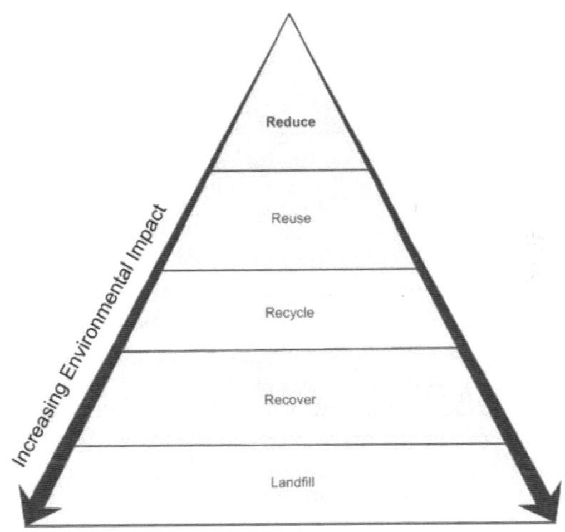

Figure 2 – Waste Hierarchy

5.3 RISKS

The main risk is to human life at the recycling and recover stages. In developing countries the controls for managing environmental and conventional safety can be notoriously lacking with injuries, deaths and the release of toxic substances all too common. But it is not solely the breaker who is responsible. In light of recent legal challenges the company that sells the ship for scrap can also be held accountable. There is an expectation for those selling their ships to do so in a responsible and safe manner; for navies the potential political and public relations ramifications could be even more detrimental (Vidal, J (Ref 9); Ormond, T (Ref 10.)

By planning for recycling the navy can make a conscious decision to be environmentally responsible by reducing the use of non-renewable resources either by removing them completely or substituting with sustainable alternatives. Reusing ships by designing them with modularisation in mind can make them much more attractive to potential buyers or reduce the logistics required to repurpose these ships for other roles.

5.4 PLASTICS

Plastic pollution has become a greater concern over recent years, with many studies revealing the effect of pollution on although we can estimate the direct impact in the short term due to mammals, birds and fish being killed by plastics it is very hard to understand exactly what the long term impact of plastics will be. One area notable area of research is to look at the effect of bioaccumulation of microplastics within the food chain and the potential impact this could have.

Microplastics pose a unique challenge due to their size they can be extremely hard to contain and clear up and often find their way into the food chain, there are primary and secondary microplastics. Primary microplastics are those which are manufactured, they may be items such as microbeads or nurdles (pellets used in manufacturing). Secondary microplastics are formed by the degradation of larger plastics though a combination of exposure to sunlight and ocean motion causing the plastics to fragment into ever smaller pieces.

The low density of most plastics allows them to float, this combined with the fact that many plastics will break down into smaller pieces but not fully biodegrade causes a build-up of microplastics within the ocean. Swallowed plastics do not digest and can build up within organisms either from directly eating plastic or from eating other animals containing plastic. Microplastics have also been found in fresh water supplies which can directly affect ship's personnel (Ref 11).

Currently the Great North Pacific Garbage Patch (Figure 3) contains over 80,000 tonnes of plastic (Ref 11).

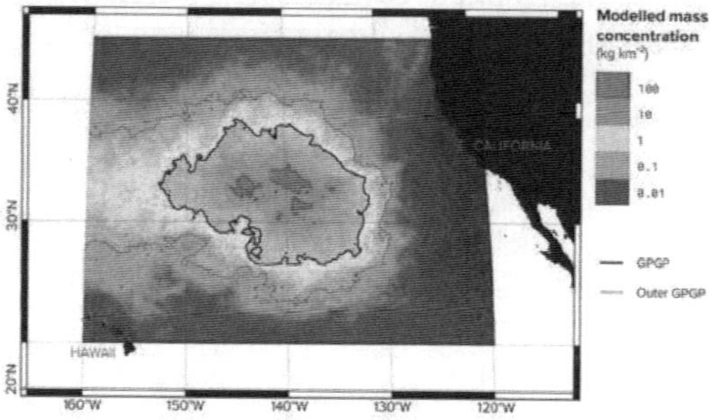

Figure 3 – The Great Garbage Patch

Warships are currently under no obligation to reduce the amount of plastic in the sea beyond the management of their own solid waste, however given the scale of the problem it is impossible to tell what regulations may come in nationally or internationally in the future, for example cruise ships wanting to visit "pristine environments" may have to directly or indirectly remove waste to mitigate for their activities.

5.5 LIVE FIRE EXERCISES

Military vessels have a requirement to test their weapons and targeting systems and this can only really be done via a live fire exercise.

However live fire exercises particularly against real targets (retired warships) presents a variety of problems as these old ships contain Polychlorinated Biphenyls (PCBs), asbestos, lead and certain other heavy metals.

In the United States (US) the military is given a permit to carry out these live fire exercises as it is regarded as essential for naval training and is considered to be a suitable approach to disposal of retired warships. The Environmental Protection Agency (EPA) which is the department within the US government tasked with protecting the environment feels that legislative reform and an increased budget for disposal is a more effective method of disposing of these old ships (Ref 12).

Other countries have removed all dangerous substances from the "target" prior to live fire exercises. The sunken hull becomes a reef which supports a variety of marine life.

Although this issue has not been as high on the agenda as the other trends, it is a particular issue unique to warships and can affect testing of weapons and training, which are two important activities a warship needs to undertake. The approach to live fire exercises and their effects on the environment varies from country to country which suggests that a single approach that is broadly acceptable to all parties may be required.

6. WARSHIP DESIGN AND CONSTRUCTION

In making a warship green and mission capable we have to consider how warships and designed and built.
It takes between 12 and 20 years from the date that a government states that a new warship is required to entry into service. In this time period many things can change. David Peer states in his paper (Ref 13) about the Canadian Patrol Frigate: "For example, the government originally envisioned a 24-ship project but it ended with 12 ships. As well, the frigates were designed for the Cold War, but between the commissioning of the first ship in 1988 and the last ship in 1996, the Cold War had ended and the exact role of the ships was evolving."

David Peer does not mention environmental and energy regulations but these are changing in parallel with the evolving role of the vessel.

6.1 DESIGNING A WARSHIP

The design spiral (Figure 4) is a traditional method of showing how a ship is designed. Environmental and Energy Efficiency requirements are shown as a green box as they need to be considered during the development of the ship's design. The equipment required for environmental protection may add weight

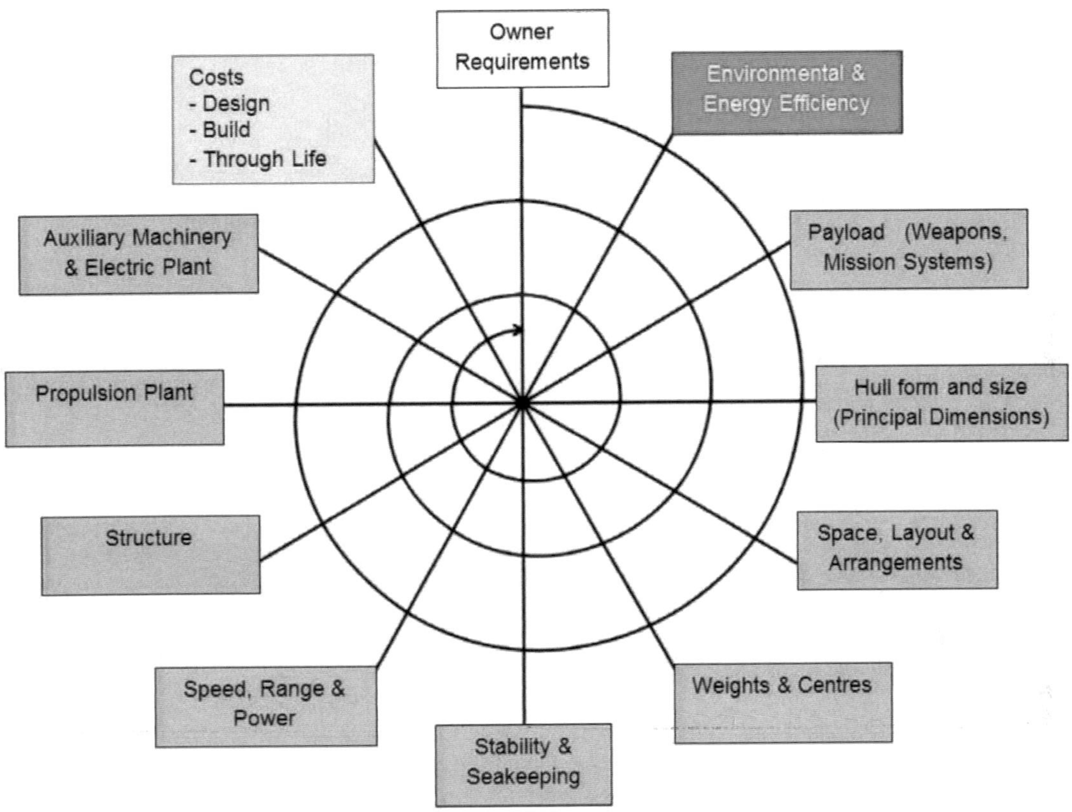

Figure 4 – Design Spiral

The yellow coloured box of "Costs" which includes Design, Build and Through Life Support is affected by these supplemental regulations and needs to be considered during the design phase.

6.2 SHIP'S LIFE OVERVIEW

The simplified Gantt chart shown in Figure 5 below shows the life of a typical warship. At the top the CADMID cycle is shown. CADMID is broadly similar to many other project lifecycles and covers Concept, Assessment, Demonstration, Manufacture, In-service and Disposal and is used by the United Kingdom MoD (Ref 14). It should be noted that the In-service phase of the cycle is the longest and exceeds the Design phase by nearly 3 times. With life extensions to vessels this could become even longer.

The Ship Lifecycle is shown in the bottom half of the diagram. For the purposes of this example the vessel's in-service life has been assumed to be 35 years with a design and build phase of approximately 14 years both of which may be substantially different in the future. The Design phase is considered as one component although the CADMID cycle breaks it up into 3 distinct areas. In reality the Design phase will consist of many option studies as well as the detailed design of the vessel. The Gantt chart also shows that there will be Major Refit Periods every 10 years which follow Rule updates. The environmental and energy reduction upgrades will be in addition to any other upgrades that a warship may be undertaking (e.g. new weapons, combat systems or propulsion equipment).

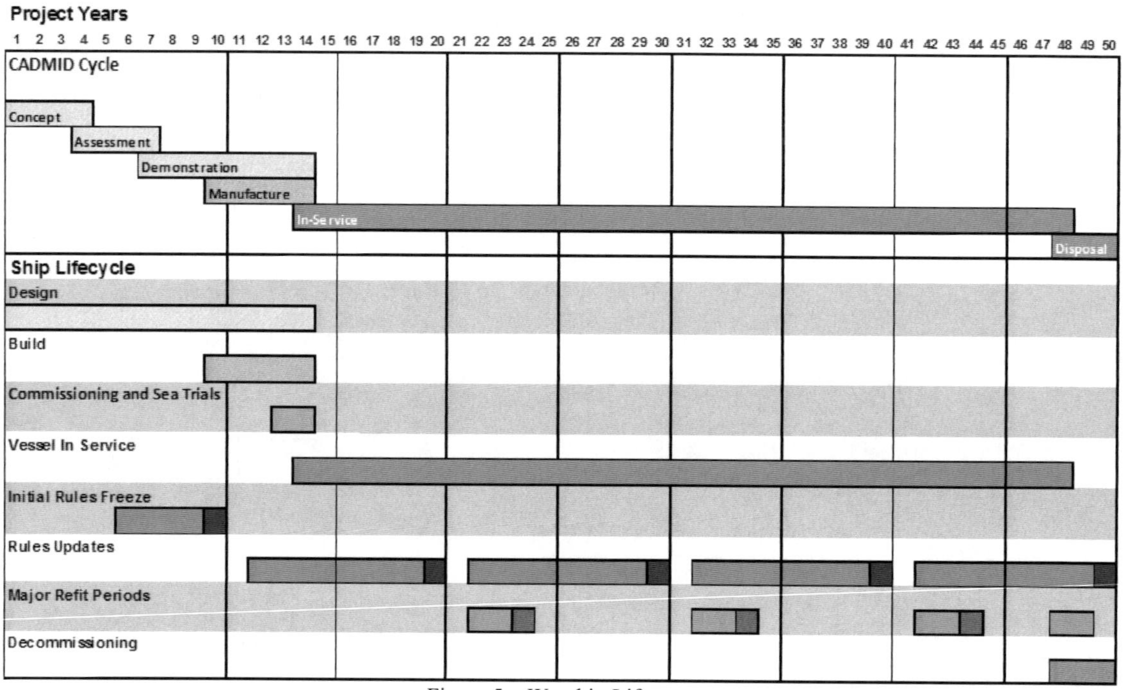

Figure 5 – Warship Lifespan

The rules changes are shown as a single square with a "tail" to the left. This "tail" is the period when rule changes are being considered and when a "watching brief" must be carried out. During the period of the "watching brief" various studies and investigations would be undertaken to determine the consequences of forthcoming changes in the regulations.

Following the rule change (when the new regulations are officially issued and brought into force) there would be a period of assessment and planning for the Major Refit (the "tail" shown in pink). The red square shows the refit year following 2 years of planning.
The ratio of time in-service to the design and build of the ship (3 to 1 possibly up to 4 to 1) highlights the need for facilitating upgrades. In the past it has been assumed that major equipment changes will only be undertaken in rare cases such as in the event of a catastrophic failure of an engine (as an example). A core assumption for new warships is that equipment will be changed throughout the life of the ship with particular attention paid to environmental protection and energy reduction equipment.

Typically emissions and energy consumption have not been a driver for upgrades but it is likely in the future that they will. Warships have not used heat recovery systems like exhaust gas boilers or jacket water distillers but they may be obliged to do so. The new regulations may be as extreme as requiring zero CO2 emissions. Battery banks and solar panels have not been used on warships but these may be an option in the future.

Some equipment may contain materials which will be banned in the future. This equipment may have to be replaced.

Equipment that prevents pollution may have to be upgraded in the future. Sewage plants, oily water separators or ballast water treatment systems may have to be replaced at intervals in the life of the ship.

6.3 FACILITATING UPGRADES

All of the above points to the need to design a warship with one eye focussed on environmental and energy regulation changes. While it is not possible to predict every possible regulatory change the ship must be designed with this in mind.

Oversized hatches are a simple solution to permit replacement of larger components. Designing engine rooms so that there is a clear route to lift out a diesel engine and replace it is more complex but is well within the realms of the possible. Modularisation is another solution although it is mainly being considered for mission modules which contain a specific set of equipment in a container. It is however possible to have engine room equipment "modularised" to facilitate changing it. Manufacturers of fuel separators already provide a separator with control panel, heater and tank all as one module which can be easily craned into (or out of) the machinery space.

6.4 UPGRADING AND UPDATING DIFFICULTIES

There are certain areas that are particularly difficult to upgrade as they don't involve a single component and they are embedded throughout the ship. These include:

a. Cables;
b. Insulation.
c. Control / surveillance systems

Cables have a particular issue that they are laid one on top of each other or tied together in groups. Often ships that have received an upgrade that requires new cabling will leave the old cables in place as it is too difficult to remove the old ones. This can create problems if the cable insulation material is subsequently banned.

Insulation is applied to the internal surfaces immediately following painting and as a result is positioned behind pipes, vent ducts and machinery. When asbestos was banned many older ships were "grandfathered" meaning that they did not have to remove the asbestos providing they did not disturb it. In the future this may change. If a material is determined to pose a risk to human life, future legislation may require that the insulation containing the material be removed as soon as it is possible to dry dock the ship.

7. NAVAL SHIP ENVIRONMENTAL COMPLIANCE (UK)

The Royal Navy is committed to having its fleet compliant with all legislative requirements as stated by the First Sea Lord Admiral Philip Jones (Ref 10). An excerpt from his environmental protection statement is quoted below.

"I am committed to ensuring that our management of environmental protection can deliver, without weakening our operational resolve, our legislative and regulatory compliance obligations, minimise harm to the environment from our actions and manage the risks which the changing natural environment and its regulation poses to our operational capability."

This says that all ships in the Royal Navy shall be both green and mission capable. Western navies that have signed up to the Naval Ship Code will likely have similar requirements but these were not investigated as part of this report.

7.1 OVERVIEW OF THE NAVAL SHIP CODE

The most well-known change in regulations pertaining to warships is the use of classification societies and their regulations being applied to warships. This change began in the 2000's and is now fully part of all future warship building. Warships still retain their own rules under the Naval Ship Code which contains the Allied Naval Engineering Publication (ANEP-77) (Ref 2) because of their unique requirement of entering areas of armed conflict which require weapon systems and additional levels of redundancy and subdivision for the ship. Under the Naval Ship Code the following certificates are required:

- Fire;
- Explosives;
- Propulsion & Manoeuvring Systems;
- Aviation (CSA).

In addition a Naval Ship Safety Certificate (NSSC) is required which covers:

- Structural Strength;
- Stability;
- Escape, Evacuation & Rescue.

The classification society regulations are used to support the above certification. As one example, the class rules on hull construction will be used to demonstrate that the ship meets the requirements of Structural Strength.
The Royal Navy has chosen to carry out a greater degree of assessment to ensure that vessels in its fleet are compliant with the current legislation. Each class of vessel has an Environmental Impact Assessment Review which covers the following areas of legislation:

1. Air Emissions;
2. Ballast Water;
3. Bilge Water;
4. De-commissioning;
5. Fire;
6. Food Waste / International Catering Waste;
7. Grey Water / Black Water;
8. Hazardous Waste;
9. Liquid Waste (other);
10. Noise - Ships Systems / Sonar;
11. Oil / Fuel Spills;
12. Ozone Depleting Substances;
13. Radiation - Equipment Containing Radiation Sources;
14. Use of Non-Renewable Fuel Sources.

The requirements of the Inventory of Hazardous Materials (IHM) are treated separately from the above document but are considered in the Review.

Each of the above is assessed according to the following 2 criteria: Impact Severity and Probability. The following tables show the definitions of each level.

Score	Severity	Definition
1	Negligible	Re-use of material, or negligible use of renewable or non-renewable resources. Produces inert waste. Negligible environmental impact. For example, temporary disturbance of common species only.
2	Minor	Low to medium use of renewable resources or low use of non-renewable resources. Non-hazardous waste produced and recycled or small amounts disposed of. Notable but limited environmental impact, negligible but widespread. For example, temporary damage to habitat of common species only.
3	Noticeable	Notable to large use of renewable resources, notable use of non-renewable resources. Notable non-hazardous waste disposal, hazardous waste (in Scotland, special waste) recycled, small amounts of hazardous waste disposal. Environmental impact limited to a small area, or widespread impact with minimal lasting damage. For example, permanent damage to habitat of common species only.
4	Serious	Significant use of non-renewable resources, limited use of toxic substances. Notable amount of hazardous waste produced. Notable lasting environmental damage. For example, destruction of habitat of common species or temporary damage to habitat of endangered species.
5	Critical	Large scale use of non-renewable resources, significant use of toxic substances. Large amount of hazardous waste produced. Large scale environmental damage with national significance, e.g. release of gases contributing to acid rain (NOx, SOx), or permanent damage to habitat of endangered species.
6	Catastrophic	Large scale use of very scarce resources or toxic resources e.g. use of heavy metals. Very large amount of hazardous waste produced. Severe widespread irreversible environmental damage of international significance e.g. release of greenhouse gases, release of ozone depleting substances or destruction of habitat of endangered species.

Table 1 – Environmental Impact Severity Definitions

Score	Likelihood Category	Numerical Definition
1	Occurs rarely, short duration	Occurs rarely, short duration
2	Annually	0 - 5 hours
3	Monthly	5 - 50 hours
4	Weekly	50 - 500 hours
5	Daily	Over 500 hours
6	Continuously	Continuously

Table 2 – Environmental Impact Frequency Definition

As part of the assessment Table 3 links Table 2 and Table 1 to highlight areas that need special consideration. The objective is to have all items within the Low Priority category and with the lowest score possible. Any items that appear in the Medium or High Priority areas will have to be justified.

Severity Category	Rarely	Annually	Monthly	Weekly	Daily	Continuously
Catastrophic	6	12	18	24	30	36
Critical	5	10	15	20	25	30
Serious	4	8	12	16	20	24
Noticeable	3	6	9	12	15	18
Minor	2	4	6	8	10	12
Negligible	1	2	3	4	5	6

Frequency Category

C	Low Priority
B	Medium Priority
A	High Priority

Table 3 – Example of a Score System

Environmental management has not failed in any way if environmental impacts cannot be removed or mitigated against. Providing there is an operational case that outweighs the environmental issues, the project can still proceed provided that the assessment has been conducted and provides evidence that the environmental impacts have been considered. This will demonstrate that practical and reasonable measures have been put in place to eliminate all but the unavoidable adverse environmental impacts. It is here where there may be benefits from managing 'Normal' impacts separately to the abnormal and emergency events that may be experienced In-Service. Care must be taken to identify aspects that can be controlled by procurement Project Teams, and full recording and justification of decisions must be carefully described. A suitable example would be the use of weapons. Weapons would not normally be discharged in an area where there is a high risk to marine life but this could be overridden by operational requirements (e.g. if the vessel must protect itself).

8. EXAMPLE OF A COMPLIANCE SYSTEM

8.1 ENSURING VESSEL COMPLIANCE THROUGH LIFE

An important aspect to meeting the environmental and energy regulations is that they must be managed through life.

Commercial vessels are required to have a Ships Energy Efficiency Management Plan (SEEMP) (Ref 18) to demonstrate that the ship is being operated as efficiently as possible. Some of the requirements for a SEEMP include:

a. It contains energy improvement measures identified by the ship owner;
b. It is ship specific;
c. It should be linked to the ship owner's broader energy management policy;
d. It must be developed in conjunction with the shore management team or company.
e. It must be reviewed and updated regularly;
f. It must involve personnel (including training and awareness development).

It is conceivable that the SEEMP will become obligatory for warships in the future. The IMO initially brought in the SEEMP as voluntary but it became obligatory in 2013.

Similarly the rules for the Inventory of Hazardous Materials (IHM) will become obligatory in 2020. Although the Hong Kong Convention has yet to be ratified, its requirements have been incorporated into European Union (EU) law by Regulation (EU) 1257/2013 on Ship Recycling. Whilst this regulation does not apply to naval warships by intent, Ministry of Defence (MOD) policy is to comply with the regulation 'so far as reasonably practicable, at least as good as those required by UK legislation.' (Ref 15).

All of the environmental protection aspects will have to be managed in a similar way. The particular problem to contend with is that the designers, ship's staff and shore management will need a joined up approach to make the system function.

At the beginning of vessel life a list of all the equipment and their relevant sources of pollution would be compiled. An example of such a list is shown in Table below. The method of compliance is also shown. The methods of compliance shown are strictly examples although they have been used on civilian and military vessels before.

It is likely that a similar justification or explanation document will be required for the installation of anti-pollution, energy reducing equipment. Even insulation may require a justification.

Table 4 below outlines the various components of the ship and their related pollutants. The column that has "Continuous" vs "Single Pollutant" is to distinguish between those pollutants such as exhaust gas which are continuous (when the engine is running) vs heavy metals which are built into a radar and would only have to be considered when the radar or certain components of the radar are replaced. LOP Review refers to Local Operating Procedures or activities that are carried out on board ship to maintain or operate the equipment.

Table 4 represents the beginning of an Environmental Management Plan (EMP). This plan needs to be commenced during the design phase of the ship so that the method of meeting the regulations is incorporated into the ship design. The preliminary EMP must consider through life issues as well as those for the initial build. Features such as access routes to facilitate equipment replacement will add to the purchase price but it is very difficult if not impossible to retrofit such features on an existing vessel. Lack of access routes will certainly add time to future refits so it is essential that the best plans be incorporated as early as possible into the design.

System	Pollutants	Continuous/ Single Pollutant	Method of Compliance at Build (start of life)	Through life management	Major Refit Requirements
Ballast	Aquatic organisms	Continuous	Ballast water treatment system	LOP Review	Replace unit
Cooling water	Energy loss	Continuous	Approved low energy system SEEMP	Reviews of SEEMP	Modernise equipment
Diesel Generators	Exhaust gas (CO_2 and NOx) Energy Loss	Continuous	Tier 2 engines with SCR Compliance document SEEMP	Reviews of SEEMP	Upgrade or replace engines
Galley	Food waste	Continuous	Suitable storage	LOP Review	
Paint	Chemical components	Continuous	Compliance document	Reviews of paints	Replace all hull paint
Propellers & Rudders	Noise Petroleum Products	Continuous	Low noise propellers LOP defining safe areas Environmentally safe oils	Testing for noise and pollutants	Change seals Use new oils (better for the environment)
Sea chests	Aquatic organisms	Continuous	Marine Growth Protection System	Testing of plant	Replace anodes
Sewage	Black water Grey water	Continuous	Marine Bio-Reactor Plant	Testing of plant	Upgrade plant
Solid waste	Packaging	Continuous	Processing and storage facility	LOP Review	New equipment
Sonar	High frequency sound	Continuous	LOP defining safe areas	LOP Review	New equipment
Ventilation	Energy losses Noise	Continuous	Approved low energy / low noise system SEEMP	SEEMP Review	New equipment
Weapons	Noise Heavy metals Chemical discharges	Continuous	EIA Review	Testing for noise and pollutants. Manufacturer to upgrade	New equipment
Bilge system	Petroleum products	Single	Oily Water Separator	Testing of plant	New equipment
Bunkering	Hose leaks	Single	Save-alls	Testing of plant	New equipment
Insulation & Cables	Materials may be restricted in the future	Single	IHM record	Manufacturer to upgrade	Replace
Radar	Heavy metals Radhaz	Single Continuous	IHM record LOP defining use	Manufacturer to upgrade	New equipment

Table 4 - Components of a Ship and Their Related Pollutants

9. CONCLUSIONS

The accelerating pace of change both in the earth's climate and in legislation is going to be an ongoing issue for warship designers. New vessel designs will have to be modular so that they can have old components swapped out and new ones installed without major work. Although this paper has not touched on it, the energy required to manufacture steel and other materials may become a consideration. It may be therefore cost effective to refurbish ships for many more years than we are seeing even at this time. The direction of legislation may change radically. The video footage of a baby sperm whale caught in plastic in "Blue Planet 2" (Ref 17) and the amount of plastics covering the oceans has suddenly come to the top of the agenda. Future warships may have to carry equipment to "clean" the oceans that they travel through. Those vessels that travel under the ocean may find that they have to do something similar. What could be the next legislation? And where will it come from? Equally where will the next conflict or natural disaster occur? All of these questions highlight the need for the warship designer to not only study, evaluate and incorporate new shipbuilding technologies but carry out similar activities on making the future warship as "green" as possible. This work will not only include a watching brief on worldwide environmental and energy legislation but may require research and development (R&D) to determine the best technology to use today and in the future.

10. ACKNOWLEDGEMENTS

The generous assistance of the members of the Safety Engineering Group at Babcock International (Devonport) is appreciated.

11. REFERENCES

1. GOVERNMENT OF CANADA website, 'Marine Acts and Regulations (TP 14609-2),' *https://www.tc.gc.ca/eng/marinesafety/tp-tp14609-2-marine-acts-regulations-617.htm,* Accessed on 11 June 2018.
2. NATO, 'Naval Ship Code', Rev G, NATO, 2017
3. IMO, 'Low Sulphur Fuel Oil Requirement,' http://www.imo.org/en/mediacentre/pressbriefings/pages/mepc-70-2020sulphur.aspx, Accessed 17 August 2018
4. CATHELCO Website, 'Marine Growth prevention systems,' *http://www.cathelco.com/mgps-overview/how-a-marine-growth-prevention-system-works*, Accessed on 21 June 2018
5. LYSTER, C., DAWE, T., 'Matching the propulsion system to the mission to reduce maintenance costs for OPVs', *Engine As A Weapon*, June 2015.
6. IMO, '2015 GUIDELINES FOR THE DEVELOPMENT OF THE INVENTORY OF HAZARDOUS MATERIALS', *MEPC*, 2015
7. CAMBRIDGE DICTIONARY, 'Definition of "sustainability" '*https://dictionary.cambridge.org/dictionary/english/sustainability*] Accessed on 8 August 2018.
8. MIZANUR, S.M., et al, Life cycle assessment of steel in the ship recycling industry in Bangladesh, *Journal of Cleaner Production,* 2016
9. VIDAL, J., 'This is the world's cheapest place to scrap ships' – but in Chittagong, it's people who pay the price', *https://www.theguardian.com/global-development/2017/dec/02/chittagong-shipbreaking-yards-legal-fight,* Accessed 10 August 2018
10. ORMOND, T., Hong Kong Convention and EU Ship Recycling Regulation: Can they change bad industrial practices soon? *Eleni Review, (2)* 2012.
11. 'The Great Pacific Garbage Patch', *https://www.theoceancleanup.com/great-pacific-garbage-patch/*, Accessed 09 August 2018
12. ZIPPEL, L, 'A Sinking Ship: EPA Regulation Of The Navy Training Program SINKEX Under The Ocean Dumping Act And The Toxic Substances Control Act,' *Washington Journal of Environmental Law & Policy*, 2013.
13. PEER, D. Cdr, 'Realistic Timeframes for Designing and Building Ships,' *Canadian Naval Review volume 9, number 1* (2013)
14. UK GOVERNMENT WEBSITE, 'CADMID Cycle UK MoD definition,' https://www.asems.mod.uk/guidance/manual/acquisition-lifecycle, Accessed on 8 August 2018
15. JONES, P. Admiral, *Environmental Protection Statement, First Sea Lord*, 2017.
16. LI JINGYI, LIU HUIHUI, CHEN J. PAUL, 'Microplastics in freshwater systems: A review on occurrence, environmental effects, and methods for microplastics detection,' *Department of Civil and Environmental Engineering, National University of Singapore,* 2018.
17. BBC, Blue Planet 2, 'Episode of sperm whale calf with plastic bucket stuck in its mouth,' https://www.radiotimes.com/news/tv/2018-01-24/blue-planet-ii-producer-explains-what-happened-next-with-that-baby-sperm-whale-sometimes-youve-got-to-step-in/, Accessed on 11 June 2018.
18. LLOYD'S REGISTER, 'Implementing a Ship Energy Efficiency Management Plan (SEEMP),' *Lloyd's Register,* 2012

12. AUTHORS BIOGRAPHY

Cody Lyster is Principal Marine Engineer at Babcock International in Plymouth. He is responsible for managing technical improvements to RN submarines and surface ships. His previous experience includes Design Manager of the Irish Offshore Patrol Vessel where he ensured compliance with classification society rules and military requirements.

Ashley James is a Safety and Environmental engineer at Babcock International in Plymouth. He provides oversight for design changes on naval surface ships and submarines; he is also analysing environmental compliance on future Royal Fleet Auxiliary (RFA) ships.

DESIGNING TO SUPPORT THROUGH LIFE ASSURANCE: A WELDING AND NDT PERSPECTIVE

J Clarke, M Sansom and **V Williams,** Babcock International Group, UK

SUMMARY

Modern Warships are expensive assets to build and maintain; this is exacerbated by an uncertain scope of operations governed by a changing geopolitical climate. Throughout the life of these assets, there is a requirement for continued assurance, with the potential for this to extend as they are sold on to other navies. Meeting these requirements is becoming increasingly challenging against a backdrop of reduced funding and a skills gap in STEM (science, technology, engineering, maths) disciplines. It is therefore necessary to invest in technologies during the design phase of projects – where modifications have a much lower cost compared to those undertaken in service. This paper will review how advancements in Welding, Non-Destructive Testing (NDT) and Engineering Assessment can be incorporated by designers to enable this demand to be met. To aid future welding productivity, certain design and build activities can be adopted which will give better access and scope to the vessel. Similarly, by engaging with future NDT trends, designers can allow for newer inspection techniques to be used. The paper will also explore how a comprehensive engineering data management strategy alleviates some of the challenges associated with through life assurance and how, in its own right, it can be an exportable product alongside the asset.

NOMENCLATURE

AP	Application Protocol
ALARP	As Low As Reasonably Practicable
CAD	Computer Aided Design
ECA	Engineering Critical Assessment
EDM	Engineering Data Management
GMAW	Gas Metal Arc Welding
LR	Lloyd's Register
SMAW	Shielded Metal Arc Welding
NDT	Non-Destructive Testing
WAAM	Wire Arc Additive Manufacture

1. INTRODUCTION

There is the potential for technology to revolutionise the way that future Warships are maintained through life. However, it is the view of the authors that to achieve increased availability, capability and reduced costs through life, maintenance must be considered during the design and build process.

Within a typical through life frigate refit package, the structural aspects are expected to account for 30% of total costs. While taking advantage of process developments has the potential for cost reduction, it is suggested the greatest savings require a change during the design phase, particularly with regards to data recording and its subsequent availability for the lifetime of the vessel or class of vessel.

The concept of design optimisation for the required application is a well understood concept within Systems Engineering, Figure 1. Various values are quoted, but for complex systems it is suggested that the cost of change in service is 29-1500 times the cost of that enacted during requirements capture and design [1].

This paper details how design changes and data capture during build can result in improvements, with a specific focus on welding process optimisation, acceptance criteria development, NDT inspection and the implementation of new inspection techniques. It provides a framework for the data that needs to be captured and managed, with a view to supporting through life maintenance and modification at reduced cost.

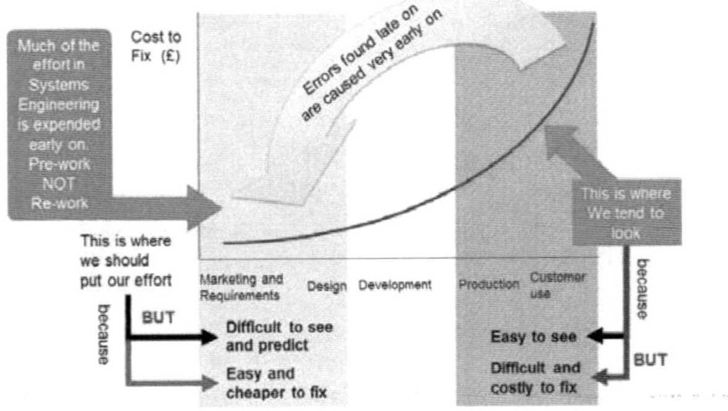

Figure 1: The comparative cost of activities to resolve issues through a product life [2]

2. NAVAL DESIGN STANDARDS

Naval Design Standards stipulate the rules for the design and construction of a naval vessel and have trends which allow for ships to adapt and integrate to different demands. Examples of trends are:

- Adaptation for advancing computer technologies
- Enhanced stealth
- Changing threats
- Increased speed
- Reinforced hulls
- Materials
- Ship building in international countries

In the UK, the Ministry of Defence governs which Naval Design Standards are used in its surface ship strategy.

Lloyd's Register (LR) Rules are often used in the fabrication of Warships and these will state the acceptable methods and standards which must be used during design and fabrication (including the above trends). However, these standards mainly concentrate on military capability, impact survivability and environmental and operational loads. They do not include designing for future maintenance, specifically for the integration of new Welding and NDT methods [3].

3. WELDING

Welding is the main fabrication process used in the construction of marine vessels. In most instances automated and mechanised welding can be used during the original build because the joints are made from new material which can be machined to the tight tolerances required of this type of welding. Additionally, the joints can be of a set length and dimension allowing the welding equipment to be planned and set-up more quickly.

When a ship docks for maintenance, original welds are removed for the replacement of equipment or for repair. When re-inserting the new components welding is normally completed manually. There are multiple reasons which cause the welding environment to change, meaning the original processes cannot be utilised again:

- Welds are of bespoke geometry in order to minimise the amount of welding required;
- There is often work-in-way which prevents the automated equipment from fitting into the space;
- New material may be attached to older (and usually of unknown cast) material causing cast-to-cast variation which can cause additional defects when using automated equipment.

In order to use mechanisation, these aspects need to be investigated and their use integrated at the design phase of a vessel's lifecycle.

3.1 MANUAL WELDING OPTIMISATION

Shielded Metal Arc Welding (SMAW) is often used in confined spaces as the process is self-shielded and gas release is easily controlled. However, this is a relatively low productivity process. During design phases, tank access should include provision of adequate venting to allow the safe use of gas shield processes such as Gas Metal Arc Welding (GMAW). This relatively simple change has the potential to increase welding deposition rates by a factor of three.

Productivity can be further improved by the use of mechanisation and ultimately automation.

3.2 REMOVING THE WELDER

3.2 (a) Mechanised Welding

Welding that requires manual adjustment of the arc can be navigated by a mechanical device [4]. Typical systems include a carriage which traverses a guide or runs on a track, see Figure 2.

Typically, these systems have a number of advantages in that they increase weld deposition rates, operators are more easily trained than manual welders and – provided welding parameters are optimised – can produce welds with very low defect rates.

Figure 2: Example of a mechanised cutting and welding carriage which utilises a track of guidance

3.2 (b) Automated and Robotic Welding

This process is similar to mechanised welding, however, the welding arc is traversed along a pre-programmed route, typically various feedback mechanisms are used to ensure the quality of the weld so that operator intervention is not required. Equipment can take various forms, however, anthropometric arms are common as shown in Figure 3.

Figure 3: Example of an anthropometric welding robot used during ship build [5]

Again, advantages include high deposition rates, high quality welds and reduced resource requirements.

3.2 (c) Design Modifications

While automation and mechanisation are used extensively during new build, their use has been limited during through life maintenance of complex Warships. Each application can be restricted in its use because of new equipment configuration for each weld and the need to increase the requirement to remove work-in-way.

Thus, if areas such as soft patches or machinery access points are designed to allow space and configuration of mechanised or automated processes, it would remove access issues and configurations can be developed prior to the docking period.

The welds produced by automation and mechanisation tend to have very low defect rates. Based on demonstrated success of limited flaws, a number of standards allow a reduction in inspection frequency and amount, again reducing overall weld costs.

Currently, within the UK there is a shortage of welders who have the skills required to support maintenance of complex Warships. Welding skill development requires significant training at high cost, whilst automated and mechanised welding operators have a different skillset, which is more easily developed through a shorter duration training package.

Similarly, if insert geometry can be standardised as part of a design, standard jigs can be produced which can greatly improve the efficiency of equipment set-up.

The potential gains are application specific. However, as a minimum, it is expected that weld deposition rates will increase by a factor of two, with much greater gains potentially possible. Further gains can be expected if builders can increase commonality between vessels of the same class.

3.3 STANDARDISATION ACROSS CLASS

While modular additions are increasingly being recognised as a cost-effective way of increasing capabilities, vessels currently in service are known to have significant non-functional variation in structures and systems due to alterations during the build process.

The result of this is that any modification has to be designed and built bespoke for each vessel. If non-functional variation could be reduced or eliminated, it will reduce the cost and complexity of class upgrades and modifications.

Standardisation also allows fabrication work to be completed while the vessel is at sea, increasing platform availability. It also allows structural elements to be transferred more readily between platforms (adding the introduction of modular operation) and thus increasing capability.

3.4 ADDITIVE MANUFACTURE

Various processes for additive manufacture have been developed, with wire arc additive manufacture (WAAM) currently being developed to produce large (> 10 kg) components [6].

Currently WAAM is most economically effective for the production of large machined components, where savings are made by reducing material wastage [7]. It is expected that there are benefits in the production of obsolescent components or those with long lead times – such as castings – that might otherwise delay in service refit or repair activities.

There are a number of technical challenges regarding the validation of material and component properties. One of the greatest challenges for the use of WAAM to produce parts for in-service is ownership of design and intellectual property. If critical components can be identified during design or the initial procurement cycle, there will be leverage to agree future design rights – something not easily obtained at the refit stage.

4. WELD ACCEPTANCE CRITERIA

Welds made during construction of a Warship are typically accepted to standards using workmanship criteria, based on the defects it is reasonable to expect a competent welder to produce. If defects are found in a weld which are outside the acceptance criteria, the defect must be removed and the weld repaired.

Repairing sections on welds can often lead to more defects being added to the weld, causing the process to be repeated until the defects are all removed. The cost of re-work on welds can often triple the original cost of the weld and has the potential to cost ten times as much [8], especially if more than one repair has to be completed or the weld repair is complex, rendering weld repair unfavourable.

4.1 ENGINEERING ASSESSMENT

Engineering assessment removes the use of workmanship acceptance criteria and instead analyses: material properties, geometry of the defect and stresses (Figure 4).

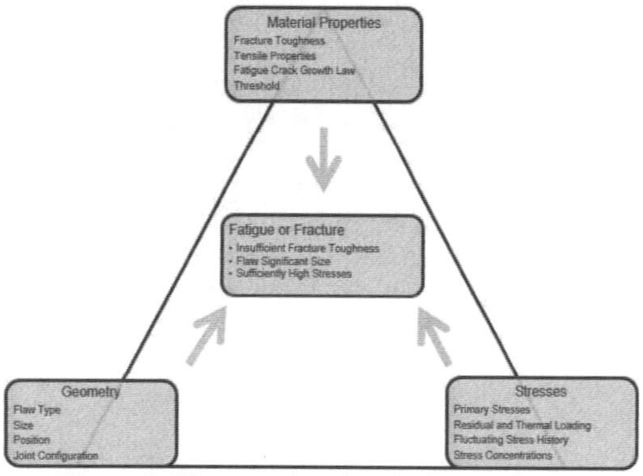

Figure 4: Diagram showing the interaction of stresses, geometry and material properties which all contribute to potential fatigue or fracture failure

Conducting a more thorough investigation of the defect considers the working environment of the weld and produces a "fitness-for-purpose" report. The "fitness-for-purpose" philosophy is that:

"A flaw is acceptable provided conditions for failure are not reached in required service life"

Therefore, repair may not be required because the working conditions of the weld will not bring about failure during its design lifetime.

Engineering Assessment could be used as a supplement to conventional design practises as they do not currently consider the working conditions of weld flaws on service performance. Extra data would need to be recorded during the design and fabrication of the vessel so that material properties could be found and used later to conduct "fitness-for-purpose" assessments. Limiting the amount of weld repair conducted during refit and repair could help:

- Relax acceptance criteria and subsequently avoid repair;
- Substantiate a reduction in NDT;
- Justify deferring repair forever or until repair is convenient;
- Forecasting failures associated with 'wear-out', minimising effort associated with ageing platforms as seen with the reliability 'bathtub curve'.

5. NON-DESTRUCTIVE EXAMINATION

5.1 COMPARITIVE EXAMINATIONS

Typically, in-service NDT examinations are completed either to validate welding activities or as part of survey activities.

If the examination is being completed as part of through life surveillance, the objective should be to closely recreate the original build inspection. In order to achieve this, it is imperative that original inspection parameters are recorded. Thus, near identical inspections can be completed and through life and build results can be comparatively examined. This has the advantage that any defects which have not propagated can be considered benign. If there has been growth, the programme may require optimisation of the surveillance package.

5.2 INCREASING INSPECTION SENSITIVITY

Over the past 15 years, the development and application of advanced examination techniques has increased the sensitivity and accuracy of inspections. This includes: ultrasonics utilising phased array transducers or time of flight diffraction for volumetric examinations; and surface examinations utilising eddy current arrays; and alternating current field measurement.

The prime driver for this has been the reduction of radiography due to ALARP principals; with added benefits such as the removal of exclusion areas allowing work to be completed concurrently to the examination.

Within petrochemical industries, the economic assessment of inspection has long been a consideration [9], while within the UK Defence Industry the NDT process selection and its application is currently standardised rather than specific to a structure.

5.2 (a) Cost Impact

The current methodology for development of inspections is to demonstrate that a new technique is at least as sensitive and accurate as that which it is replacing [10]; noting that a number of methodologies are covered, by current standards.

In reality, the new techniques are significantly more sensitive; meaning the examination of in-service structure highlights defects that may not have been detected or acceptable at build.

Recognising that once a defect is identified there is typically a requirement to repair, the new techniques are potentially increasing the through life repair burden with limited benefit to the structure.

Given that developments in NDT inspection capability will once again increase the sensitivity and volume of data developed by inspections i.e. full matrix capture. This is potentially an issue that is still developing.

Therefore, it is suggested that implementation of new inspection techniques includes the development of specific acceptance criteria to ensure that only critical or potentially critical defects are repaired; either using ECA for a specific flaw or by setting limits on current workmanship criteria.

6. ENGINEERING DATA MANAGEMENT

The concept of Engineering Data Management (EDM) is not a new one. T W Siddle [11] provides some background to the advent of EDM – as computers became more readily available, large engineering organisations looked to automate aspects of engineering design and manufacture. Over time, influenced by the success of these automated aspects, there was a drive to increase the scope of computerised operations to increase throughput of overall processes. As the bulk of these automated aspects utilised bespoke and application specific programs – each maintaining a range of incompatible data files – the difficulty of achieving this overall process aim was greatly exacerbated. It became apparent that the data generated by these programs must be centralised and in a compatible form with other application programs.

During the late 1970s and early 1980s, different sources developed standards for Computer Aided Design (CAD) data exchange, two chief ones being *'Digital Representation for Communication of Product Definition Data'* and *'Verband der Automobilindustrie – Flächenschnittstelle'* developed by the US National Bureau of Standards and the German Automotive Industry Association respectively. These would then be superseded by ISO 10303: Industrial automation systems and integration – Product data representation and exchange. The purpose of this standard is:

"To provide a mechanism that is capable of describing product data throughout the life cycle of a product, independent from any particular system"

and is organised in to a series of parts focusing on 'Environment', 'Integrated Data Models' and 'Top Parts'. Within 'Top Parts' the standard set is further subdivided into 'Application Protocols' (APs). These APs cover a particular application which can be roughly grouped into Design, Manufacturing and Life Cycle Support.

In the design sphere EDM is now firmly in place, with engineering data being managed in systems that are capable of transferring that information between stakeholders and companies in a readily available fashion. It still however has not fully embedded itself in manufacture and life cycle support, with many operations still running using individual databases and spreadsheets, and in some instances still on hard copy paper forms.

In the age of Industry 4.0, where around 70% of value is left uncaptured due to a lack of data management processes, policies and infrastructure [12], the time is now to invest in EDM solutions in order to enable cost savings, alleviating issues seen in manufacture and through life assurance.

6.1 WELDING AND NDT DATA AVAILABILITY

One of the main inefficiencies in Welding & NDT operations is the availability of data, be it from design and manufacture or during life cycle support.

An example of this can be seen with weld procedures – currently the procedures developed for the build of a vessel are not made available to those maintaining the vessel. As a result new procedures are developed, tested and qualified at the expense of both time and cost. Currently each procedure costs thousands of pounds to be developed and qualified. This comprises engineer, welder and classification society labour as well as test materials and consumables.

Another example pertains to the history of a weld. Currently it can prove challenging to understand what defects present in a weld were there at build, and how they have changed over time. If this information was made available and the engineering assessment methods explored in Section 4.1 utilised, potential repairs to build defects could be avoided as well as providing a greater insight into the structural health of the vessel.

6.2 WELDING & NDT DATA INITIATIVES

There are many initiatives that an EDM solution would enable; two chief ones being Parameter Monitoring and Condition Monitoring for Welding and NDT respectively.

6.2 (a) Parameter Monitoring

By collecting welding parameters in real time throughout both build and refit activities, and associating the data to the specific weld, welder and welding unit could allow Welding Engineers to ensure that welding was undertaken in accordance with the parameter ranges specified in the Weld Procedure as well as ensure welders were adequately qualified for the joints that they were welding.

Conversely a digital system could itself be set up such that welders would have to identify the joint, equipment, procedure and themselves at the point of welding – if the correct conditions are not satisfied then the weld is not able to take place. This could potentially reduce defect rates and costly rework operations, as well as reduce time and effort spent completing welding quality documentation – using the system to complete the documentation with the collected data.

6.2 (b) Condition Monitoring

Installing application specific ultrasonic equipment onto a vessel's structure at build, particularly in areas of suspected high corrosion, and linking them to a data collection device (cloud based or otherwise) would allow the structural health of a vessel to be better understood whilst she was at sea. Thus when the time comes for refit operations to take place, the specific structural inserts required could be identified before arrival and the sections fabricated and on the dockside ready for installation.

Not only would this have the potential to reduce programme times, minimising surveying activities to identify areas requiring structural recovery; but could also allow dockyards to level out workload peaks and troughs particularly involving skilled welding labour.

6.3 WELDING & NDT DATA EXPORTABILTY

It has been suggested that an EDM solution has the potential to enable cost savings, as well as alleviate issues seen in through life assurance. It therefore follows that should a vessel be sold on to another Navy, the data that was collected could also be sold on as a package that would reduce their through life costs. This could follow a software 'release' type model or be marketed as one discrete package depending on the needs of the customer or the way in which the data was collected.

7. CONCLUSIONS

This paper has shown that current Naval Design Standards lack detail in ensuring forward planning for the through life maintenance of Warships. The changing geopolitical climate and subsequent change in naval demands is unlikely to change in the near future so the Welding and NDT issues faced today are unlikely to disappear.

There are many design processes which could be reviewed and re-engineered so that through life costs of Warships are reduced by forward planning at the proposal stage. Integration of this knowledge into Naval Design Standards would ensure that designers are aware of this increasingly important issue and lead to the development of more ideas to be incorporated at the design stage.

One change at the design stage could be the introduction of an EDM system which affects many of the maintenance improvements mentioned in this paper. By improving the sharing of knowledge and data, more efficient technologies could be more easily implemented e.g. engineering critical assessment and new NDT examinations; better communication of data will improve efficiency and thus reduce the cost of maintaining a Warship. It is thought that a funded study into the different design methods which could be adopted to improve Welding and NDT maintenance could look into these ideas in more detail to see which areas hold the most savings.

8. REFERENCES

1. NASA, 'Error Cost Escalation Through the Project Life Cycle', *NASA Johnson Space Centre* https://ntrs.nasa.gov/archive/nasa/casi.ntrs.nasa.gov/20100036670.pdf, accessed 20/04/2018

2. BABCOCK INTERNATIONAL GROUP, 'A Systems Approach to Systems Design', *Engineering Development Program*, 2016

3. G ASHE ET AL., 'ISSC Committee V.5: Naval Ship Design', *The 16th International Ship and Offshore Structures Congress*, 2006

4. AMERICAN WELDING SOCIETY (AWS), 'Standard Welding Terms and Definitions', *AWS Committee on Definitions A3.0*, 2001

5. SHIP TECHNOLOGY, 'Image of a Welding Robot', *http://www.ship-technology.com/wp-content/uploads/sites/16/2017/10/ship-robot-l.jpg*, accessed 23/05/2018

6. S W WILLIAMS ET AL., 'Wire and Arc Additive Manufacturing', *Materials Science and Technology Volume 32*, 2016

7. Y ZHAI, 'Early Cost Estimation for Additive Manufacture', *Cranfield University*, 2012

8. G SLATER, 'The Effect of Repair Welds on Service Performance', *65th AWS Annual Meeting*, 1985

9. M WALL and F WEDGEWOOD, 'Economic Assessment of Inspection – the Inspection Value Method', *7th European Conference on Non-Destructive Testing*, 1998

10. BSI, 'PD CEN/TR 14748:2004 Non-Destructive Testing – Methodology for qualification of non-destructive tests', 2004

11. T W SIDDLE, 'Weaknesses of Commercial Data Base Management Systems in Engineering Applications', *17th Design Automation Conference*, 1980

12. MCKINSEY GLOBAL INSTITUTE, 'The Age of Analytics: Competing in a Data-Driven World', 2016

9. AUTHORS BIOGRAPHY

James Clarke works for Babcock International Group as the Business Support Manager for Welding Design & Non-Destructive Testing and is currently managing the implementation of a digital Welding Quality Management System. James studied Ship Science with Naval Architecture at the University of Southampton.

Mark Sansom is currently Chief Welding Engineer at Babcock International Group's Devonport Royal Dockyard facility. He is responsible for welding engineering capability which supports the in-service maintenance of the UK RN submarine and surface fleets. Mark studied Naval Architecture at the University of Plymouth before completing a Masters in Welding Engineering at Cranfield University. His previous roles have included development of NDT techniques and projects related to the welding of high strength steels.

Veronica Williams works for Babcock International Group as a Senior Welding Engineer involved with weld procedure development and is currently implementing the ECA process. She has studied at the University of Liverpool and Cranfield University growing her engineering knowledge and technical welding understanding and is an Incorporated Engineer.

Trends in Environmental Requirements and their Impact on the Future Surface Vessel

J A N Livingstone, BAE Systems Naval Ships, UK

SUMMARY

Whilst navies benefit from disapplication or derogation from many international and national environmental regulations, governments' policies and societal concern mean that they are now expected to meet or exceed the standards set by these environmental regulations. The role of the designer is therefore to ensure that warship design does not restrict the ability of a nation to undertake military operations without constraint. This paper reviews the history of the key environmental regulations and identifies indicators for future development. Using examples from the Type 26 Global Combat Ship it will illustrate how the design has minimised the impact of future trends. The review identified some future design requirements to manage oil, black water and grey water, and significant future design requirements for air emissions in particular carbon dioxide. Early adoption of design requirements to mitigate future environmental regulatory trends is critical in ensuring these requirements are achieved for naval ship design.

1. INTRODUCTION

Environmental awareness and the drive to control humans' impact upon the planet is not a new phenomenon restricted to the 20th Century onwards and examples dating back to the 9th Century [1] can be cited. Initially driven by the need to understand the direct impact upon human populations, such as waste management and the link to disease (for example Benjamin Franklin in the 17th Century [2]), it has increasingly been necessary to understand humans' impact upon the earth's natural capital (the world's stocks of natural assets which include geology, soil, air, water and all living things) that we rely upon for survival. This is particularly relevant since the industrial revolution introduced industrialised, mechanised and increasingly large scale production of food and goods to support populations.

In response to this, numerous environmental incidents and improved scientific understanding, environmental controls have been put in place through nation states' and international regulations and legislation. As the world's population expands, the mechanisms required for the prevention of disease and preservation of earth's natural capital are required through stricter environmental regulation and control. This environmental regulation and control is also broadening its scope from traditional end-of-pipe pollution and pollution control towards holistic through-life considerations of resource input, consumables minimisation and management. For example the 'Circular Economy: decoupling economic activity from the consumption of finite resources, and designing waste out of the system' [3].

Governments and Governmental organisations such as Defence are not immune to the controls implemented. Whilst Defence undertakes some unique activities they must also strive to comply with environmental rules and regulations as far as reasonably practicable. This is particularly important if, by not complying with environmental rules and regulations, it results in:

- A restriction on the type of operations that may be conducted
- A restriction in the geographical areas of operations
- Increased time and cost in providing temporary compliance solutions
- Increased support costs due to materials restriction and substitution
- Increased adverse public scrutiny of operations due to non-compliance with subsequent restrictions

Given the long life of its materiel this leaves Defence in the difficult position of environmental compliance against the backdrop of continually changing and tighter environmental controls and regulations. Therefore the need to anticipate and design against future environmental rules and regulations is required.

The aim of this paper is to review the history and progress of environmental requirements, namely those from legislation, regulations, policy and voluntary agreements in order to signpost the direction and timescale for change.

2. APPROACH AND SCOPE

2.1 APPROACH

It's possible to design against requirements derived from known and implemented environmental rules and regulations but to anticipate and design against future requirements is more difficult. By reviewing the rate and type of requirements implemented via rules and regulations it is possible to estimate trends and enable future ships to be either fit for or with the necessary equipments to meet those trends.

This paper reviews indicators that could signal a change in environmental rules or regulations coupled with a review of extant legislation to determine how far the rules and regulations have already progressed. The environmental protection principles underpinning these rules and regulations shall also be used to assist in evaluating any identified trend.

The following indicators shall be used in this review:

- Extant legislation

- Known forthcoming legislation
- Voluntary Agreements and Guidance
- White papers, proposal papers or guidance papers
- Environmental incidents
- Societal environmental awareness
- Geographical extent

2.2 SCOPE

The subject areas of environmental protection, environmental sustainability and associated environmental rules and regulations are very broad. This paper will focus on international environmental regulations and conventions supplemented by UK and EU regulations and legislation. The latter is chosen as the experience gained during Type 26 Global Combat Ship (Type 26) design has been influenced by UK and EU regulation; and that typically, UK and EU regulation is at the forefront of environmental protection. Moreover, the focus will be on the environmental issues covered by the MARPOL Convention.

This paper will not review the following:
- Rules and regulations that are specific to tankers and tanker operation
- Rules and regulations that do not result in design changes, for example port authority requirements
- Minor amendments that refine specific technical application of rules and regulations
- Nation state regulations that implement international conventions; it is assumed these are the same

2.2 (a) Regulatory Context

Societal awareness was included as an indicator as this is often a precursor to pressure to regulate and control; this can typically be seen following environmental incidents and accidents.

Regulatory instruments are cognisant of the availability of technology to deliver them and disapplications, derogations or authorisations (to pollute) include an assessment of technology readiness and the cost of implementation. Where technology limits the level of compliance the legislation includes regular reviews and updates.

3. POLLUTION FROM OIL

The release of petroleum and oil into the marine environment affects populations of invertebrates, birds and mammals particularly those that inhabit shallow water and the littoral zone and those that have to cross the air-sea interface [4]. Particularly sensitive habitats include polar, salt-marsh, coral reef and mangrove swamps. Wider impacts can occur to recreational amenities and the socio-economic impacts of introducing potentially carcinogenic compounds into commercial fisheries.

Regulation of marine oil pollution was established in the 1950s with the advent of an increase in global fuel consumption and the rise of bulk sea oil transport. As more and larger oil tankers were used the risk this posed in the event of an accident was recognised. It was also recognised that oil pollution came from routine ship operations [5]. OILPOL, the precursor to MARPOL, came into force in 1958 establishing prohibition for discharge of oil. The International Maritime Organisation (IMO), formerly the Inter-Governmental Maritime Consultative Organisation, met in 1959 with a focus on the safety of life at sea. Subsequent oil pollution disasters, such as: the Torrey Canyon, Cornwall in 1967; the Sea Star, Gulf of Oman 1972; Argo Merchant, U.S.A., 1976; Atlantic Empress, Tobago, 1979; and, the Amoco Cadiz, Brittany, 1978; resulted in an increase in environmental awareness, environmental activism and legislation [6]. MARPOL 73 (the International Convention for the Prevention of Pollution from Ships) was introduced with subsequent stricter controls and design requirements via MARPOL 78 which included the banning of oil in forepeak tanks.

Extensive amendments to MARPOL (Annex I Regulations for the Prevention of Pollution from Oil) were implemented in 1984 including: provisions for the control of oil discharges; operations in Special Areas; oil discharge monitoring and control systems; crude oil washing; oily-water separating equipment; pumping, piping and discharge arrangements; and, damage stability criteria. In the response to the Exxon Valdez, Alaska disaster in 1989, double hull or oil outflow assessment requirements were introduced, with single hull phase out of tankers being introduced as a result of the Erika, France disaster in 1999. This phase out timetable was altered after the Prestige, Spain disaster in 2002. Over this time, amendments to MARPOL Annex I were introduced: additional Special Areas, for example the Antarctic; a discharge limit for oil in water content of 15ppm from machinery spaces; and, the oil filtering equipment is required to be alarmed and must automatically stop discharge if in Special Areas.

After this time further MARPOL Annex I amendments were introduced implementing: protective location of fuel tanks; connection requirements regarding oil sludge connections; and. the Polar Code requirements as applied to Annex I.

In summary, at present MARPOL Annex I requires tank protection or designed arrangements; a limit on the oily bilge water that can be discharged coupled with treatment equipment requirements and a complete ban on discharge of any oily mixture in Polar regions. There are additional oily bilge water equipment requirements for Special Areas that comprise:
- Mediterranean Sea
- Baltic Sea
- Black Sea
- Red, Sea

- 'Gulfs'
- Gulf of Aden
- Polar (Antarctic and Arctic)
- North West European Waters
- Oman area of the Arabian Sea
- Southern South African waters

IMO designated Particularly Sensitive Sea Areas (PSSAs) have also been established which can require specific measures for environmental protection. These usually come in the form of restricting ship movements and creating 'areas to be avoided'. Examples of the extant PSSAs comprise: The Great Barrier Reef, Australia; The Wadden Sea, Denmark, Germany, Netherlands; The Galapagos Archipelago, Ecuador; and Tubbataha Reefs Natural Park, the Sulu Sea, Philippines.

3.1 Future Trends

The regulation of maritime oil pollution has been in force for over 60 years with progressive implementation of controls over the release of oil and oil-water mixtures into the sea. These have covered accidental release, normal release and specific geographical restrictions through to a complete ban in Polar regions. Lloyds Register® has already introduced the ENV(OW) voluntary standard which is a good indicator for future trends. This requires various design requirements including: alarmed tanks and discharge of bilge arisings at the lower 5ppm oil-in-water content requirement.

With the exception of specific port authority restrictions, there are no broad 'distance from shore' restrictions applicable to oil discharge. For environmental protection this would be a logical next step as shallow water and littoral water zones are the most sensitive environments. This could be a future restriction, for example 'global prohibition of discharge of oil or oil-water mixtures within 12 nautical miles of land'. There doesn't appear to be any move to implement such a restriction via proposed papers, working groups or societal discussion. Focus may only be on ensuring effective compliance with the current regulations.

We can also use the concept of ideality from TRIZ (the theory of inventive problem solving) [8] to assist in understanding any future trends; the ideal environmental state for the ship is for zero emissions of oil into the sea causing no oil pollution. The Polar regions are a good example of this state and require ships operating in this area to be provided with design requirements covering accidental release and tanks sized to hold all bilge arisings. Can we expect a similar zero discharge situation elsewhere and over what timescale? As already described there doesn't appear to be any discussion in implementing such measures.

The Type 26 Project has mitigated the impact from the trend of oil discharge restrictions by including the Lloyds Register® ENV(OW) as a requirement and providing bilge arisings endurance. The Type 26 design baseline and the equipment purchased includes an oily water separator that conforms to the extant Annex I requirements and the more stringent discharge standard of 5ppm oil-in-water content. To meet the ENV(OW) requirements the design has also paid particular attention to accidental discharge of oil from tank air escapes undertaking risk assessments and implementing the necessary sensing, controls and gravity fed systems.

Figure 1 summarises regulation activity and potential future trends. Whilst there doesn't appear to be any trend that may affect a change in the legislation in the short or medium term the Lloyd's Register ENV(OW) notation suggests a future restriction on oil in water content concentrations which would require a change in the oily water separator equipment being included in future designs or during refit. In the medium term this could result in a requirement for oily water separators to remove all oil from discharge. There are manufacturers providing equipments that claim to reach this standard and the approach for designers could be to ensure there is enough space to upgrade to this equipment in the future.

Given the impact upon the Arctic as a result of a changing climate, Arctic shipping routes such as the Northwest Passage or Northern Sea Route could become a viable shipping route in the medium to long term (by the 2040s) [8, 9]. This could conceivably result in an increase in operations in the Arctic region and a corresponding need to design ships that can comply with a zero oil discharge endurance requirement.

4 WATER POLLUTION

4.1 POLLUTION FROM EFFLUENT DISCHARGES

Discharge of liquid effluent from ships, whether black or grey water, impacts the marine environment through: possible introduction of microbial pathogens presenting potential health risks either directly or via filter feeding organisms; reduces the availability of oxygen for marine flora and fauna; and, it may contribute to the incidence of algal blooms [10, 11]. This is reflected in the discharge standards applied to effluent comprising thermotolerant coliforms, suspended solids, biological oxygen demand and chemical oxygen demand. Sensitive habitats are those where human exposure can occur, littoral and coastal waters and where recirculation of water bodies is reduced (for example semi-enclosed seas such as the Baltic).

Regulation of black and grey water emissions into the marine environment is principally through MARPOL Annex IV which came into force in September 2003. This restricted discharge of black water (toilets, urinals, medical facilities drainage) at certain proximities to land and under certain conditions i.e. comminuted and disinfected or after treatment. It also introduced procedural controls governing how discharge could take place.

MARPOL Annex IV has subsequently been amended to include discharge standards for sewage and implement the requirements of the Polar Code to Annex IV in addition to survey and procedural controls. In 2011 a further amendment was adopted which defined the Baltic as an Annex IV Special Area introducing grey water discharge standards that included additional phosphorous and nitrogen requirements for passenger ships. This entered into force in 2013 and subsequent amendments in 2016 confirmed the implementation dates between 2019 and 2023 to take effect.

4.1 (a) Future Trends

Figure 2 summarises regulation activity and potential future trends. The introduction of MARPOL Annex IV has protected coastal, port and harbour environments since 2003 and it has assumed that in the open seas the environment is capable of assimilating sewage discharges through break up and microbial action. The recent amendment introduced grey water discharge restrictions to passenger ships as it is recognised that they discharge large volumes of black and grey water. The Baltic is a semi-enclosed sea and subject to reduced amounts of circulation, therefore reduced dilution and dispersion of the black and grey water. The Baltic is also subject to high levels of passenger ships. Could there be a similar scenario that may require similar restrictions? The Black Sea, Red Sea, Caspian Sea and Persian Gulf are all enclosed or semi-enclosed water bodies; with the Black Sea and Red Sea receiving passenger ships these could be candidates for further restrictions in the medium term.

Other sensitive environments include the Polar regions which are subject to some additional discharge requirements. Given the sensitivity of Polar environments and previous calls for additional discharge restrictions [12] there could be additional requirements such as a complete discharge ban or mandatory treatment standards. The timeline for this implementation could be aligned with an increase in shipping in these regions [8, 9].

The Lloyd's Register ENV(GW) notation is another indicator of potential future restrictions and one which is aligned to a progressive reduction in environmental impact. This notation requires treatment of all black and grey water (or retention) and would significantly reduce the impact of discharge.

The Type 26 Project has mitigated the impact from the trend of effluent discharge restrictions by treating all black and grey water arisings. The Type 26 is achieving this through use of a membrane bioreactor (MBR) meeting the Lloyds Register® ENV(GW) discharge standards. The design includes sufficient tank volume for both black and grey water and endurance for the production of biosludge from the MBR.

In summary, given their nature both the Black Sea and Red Sea are candidates for further restrictions on effluent discharge in the short to medium term. The Polar regions could be subject to additional treatment requirements in the medium term and in the long term there could be a move towards treatment of both black and grey water. All will require design changes that range from altered treatment plant (that can meet standards including phosphorous and nitrogen), altered pipework to retain grey water on board for treatment and tanks to retain biosolids produced by such treatment plants. Alternatively, sufficient volume tanks to retain all effluent arisings should be included in design requirements.

4.2 POLLUTION FROM WASTE

Globalisation and maritime transport activity has greatly increased [13]; there has also been a greater increase in use and diversity of man-made materials, in particular synthetic materials and plastics [14]. Discharge of waste or garbage overboard was, for many years, a convenient way of disposing of ship generated waste, particularly under the assumption that the marine environment could assimilate and manage the waste. This assumption has changed as the impacts from marine waste disposal, particularly plastics, are being understood. These include wildlife entanglement, potential impacts from wildlife ingestion of plastics and the more recent focus on micro-plastics and their potential impacts [15].

In response to the awareness of the impacts from waste pollution the IMO implemented Annex V of MARPOL the Prevention of Pollution by Garbage from Ships. This came into force in 1988, establishing restrictions on garbage disposal outside special areas (no plastic, other wastes have a distance from land restriction) and inside special areas (food waste with a distance from land restriction, all other wastes prohibited). Annex V and later amendments up to 2015 established a number of Special Areas that comprise:

- the Mediterranean Sea
- the Baltic Sea
- the Black Sea
- the Red Sea
- the Gulfs
- the North Sea
- the Wider Caribbean region (including the Gulf of Mexico and the Caribbean sea)
- the Antarctic

Subsequent amendments to Annex V have implemented: procedural controls (e.g. management plans, record books, placards); and, prohibited plastic contaminated incinerator ash disposal. Significant amendments in 2011, which came into force in 2013, completely revised Annex V to prohibit the discharge of almost all types of garbage into the sea with some very limited exceptions (food or food related waste). The requirements of the Polar Code were implemented in 2015. Annex V is supported by implementation guidelines issued in 2012 which were updated in 2017.

In response to the above requirements ship design now includes the retention and management of almost all wastes generated on board. This could be simply providing enough waste storage space through to complex pyrolysis plants. Ship design also needs to consider how to manage food waste when operating in locations where it cannot be discharged; in littoral waters at certain distances from land.

4.2 (a) Future Trends

Figure 3 summarises regulation activity and potential future trends. From the late 1980s over a 20 year period regulations restricting the discharge of waste into the sea have progressively tightened now prohibiting all discharges; food waste may be discharged under certain conditions as the environment is considered capable of assimilating it.

Can we expect further tightening of Annex V given that discharge of almost all waste is already prohibited? There has been a lot of recent focus on plastic waste, for example through the recent BBC® Blue Planet Series in 2017 [16], through evidence of its proliferation into the marine environment [17, 18] and organisational response from the IMO [19] to nation state reaction [20, 21]. However, the proposals would be principally land-based and there are no indicators likely to affect ship design. The focus for shipping could come from enhancing enforcement of Annex V rather than additional nugatory restrictions.

In summary, there doesn't appear to be any trend that may affect a change in the legislation in the short to long term. Current legislation already prohibits the discharge of wastes of concern and there are no indicators to suggest otherwise.

5 AIR POLLUTION AND CLIMATE CHANGE

Ships still rely upon the combustion of fuel for power and propulsion whether via diesel engines or gas turbines. Combustion products primarily comprise NOx (oxides of nitrogen), SOx (oxides of sulphur) and particulate matter which contribute to air pollution and impact human health particularly in coastal and port populations. Whilst the contribution of ships to air pollution is relatively low it is increasing [22]. During combustion carbon dioxide is also emitted which contributes to climate change [23], particularly significant in the light of a continued upward trend in fuel consumption by the shipping sector [13] and the growing contribution of shipping to greenhouse gases [24].

The shipping sector has been slow to implement measures to mitigate emissions to atmosphere despite the issues being identified and acted upon since the 1970s. MARPOL Annex VI Prevention of Air Pollution from Ships only came into force in 2005 (although being adopted in 1997) and introduced standards on NOx and SOx emissions, volatile organic compound (VOC) emissions from cargoes, shipboard incinerator emissions and use of ozone depleting substances.

In the light of technological improvements the process of revising Annex VI was commenced in 2005, the amendments being adopted in 2008 which came into force in 2010. This introduced:

- a progressive reduction in SOx and particulate matter emissions through reductions in sulphur in fuel globally (later amendments adopted a 0.5% sulphur in fuel limit from 2020) and within Emission Control Areas (0.1% sulphur in fuel limit)
- a progressive reduction in NOx emissions standards globally (Tier II) and within Emission Control Areas (Tier III)
- amendment of the NOx Technical Code

Subsequent amendments to Annex VI established the following Emission Control Areas (ECA):

- North American area (SOx and NOx in force)
- the United States Caribbean Sea area (SOx and NOx in force)
- the Baltic Sea area (SOx in force, NOx from 2021)
- the North Sea area (SOx in force, NOx from 2021)

These amendments also introduced further technical guidelines addressing use of selective catalytic converters (SCR) which had been established as a viable solution to meeting NOx Tier III standards.

Adopted in 2011 and entering into force in 2013, amendments were made to Annex VI introducing measures to reduce emissions of CO2 through energy efficiency improvements (the Energy Efficiency Design Index) and procedural means (Ship Energy Efficiency Management Plan). This was preceded by formation of the Intergovernmental Panel on Climate Change in 1988, agreement for governmental action on climate change in the Rio Earth Summit in 1992, creation of the Kyoto Protocol (the United Nations Framework Convention on Climate Change) in 1997 and subsequently by the Paris Agreement in 2015. In 2018 amendments came into force requiring data collection for fuel consumption, prompted by the EU implementation of the Shipping Monitoring, Reporting and Verification Regulation (MRV) which already requires ships' reporting of emissions. The IMO has also adopted a 2018 strategy to reduce greenhouse gas emissions by at least 50% by 2050 compared to 2008, decarbonising the sector as soon as possible in this century.

5.1 Future Trends

Figure 4 summarises regulation activity and potential future trends. Whilst the shipping sector has been comparatively late in implementing regulations to reduce environmental impacts from air emissions much advancement has been achieved since 2005.

For air pollution from SOx there have not been any indications that emissions standards could tighten with the global sulphur in fuel requirements not implemented until 2020. Naval vessels use low sulphur F76 which already meets future global requirements and those for the stricter sulphur ECAs so no future requirements are anticipated in the short to medium term. There may be geographical changes as other locations suggested as potential ECAs include Mexico, the Mediterranean Sea, Japan and Singapore [25]. Implementation of further sulphur ECAs will not change the design requirements, just stimulate the supply of low sulphur fuels. However, there are other routes to meet the intent of the sulphur in fuel requirements that comprise [26]:

- Scrubbing engine emissions. The cost, space, weight, complex operation and support penalties are high for this option.
- Change of fuel for example liquefied natural gas (LNG) or hydrogen. The cost, safety and support penalties are high and for hydrogen the technology maturity low.

For air pollution from NOx there have not been any indications that emissions standards could tighten although geographical changes may occur should new ECAs be ratified which may include Mexico, the Mediterranean Sea, Japan and Singapore [25]. Other factors prompting a change in design would be a change in technology for meeting Tier III emissions standards. Current compliance approaches comprise [27]:

- Selective Catalytic Reduction (SCR). Reduction of NOx through a post-engine catalysed reaction dosed with urea solution. Cost, space (both catalyst and urea tanks), weight, support and penalties with shock considerations as well.
- Exhaust Gas Recirculation. Cost, weight, hazardous waste production, support and fuel use penalties.
- Alternative Fuels. Such as liquefied natural gas (LNG) with the cost, safety, power and support penalties are high.

Other approaches include: direct water injection; fuel-water emulsion; intake air humidification; and, humid air motors. These have low technology readiness levels and would not fully meet Tier III on their own. For NOx emissions it is worth noting that these have not been applied to gas turbines. The regulations are aimed at commercial shipping which rarely use gas turbines and due to the combustion process for gas turbines emissions are already close to Tier III standards.

The Lloyds Register® ENV notations closely follow the future requirements in Annex VI for SOx and NOx and do not suggest a tightening of the emissions standards in the future.

The Type 26 Project has mitigated the impact from the trend of increased NOx emissions requirements by including SCR into the diesel engine design. The Project made an early decision to adopt Tier III compliance, at that time the North American NOx ECA had been ratified, and as both the North Sea and Baltic were extant SOx ECAs it was anticipated that these areas would progress to NOx ECAs thus including Type 26 home base(s). Addition of SCR in the design required a number of considerations that included: space and weight higher up in the ship – each engine requires an SCR; an SCR bypass system as the catalyst is friable and could fall into the engines during a shock event; urea tanks with enough volume to meet the required endurance; auxiliary systems to deliver the urea solution to the SCR; and, a urea dosing control and monitoring system. Early adoption was key to ensuring all of these requirements could be included in the design. The Project will be demonstrating it has met the standards through the Lloyds Register® ENV(NOx-3) notation.

One aspect of air emissions that is not directly regulated at present is black carbon and particulate emissions. Produced by incomplete combustion and / or fuel impurities (e.g. sulphur in fuel) these can have both climate warming and cooling effects and health impacts as well. This has been discussed at the IMO with a formal definition of black carbon being agreed in 2015. Based on health impacts, particulate regulation is most likely to occur in already established ECAs whilst black carbon impacts would be regulated in Polar regions. There is no indication of timescales for implementation at present. Future measures that could come into force to address black carbon and particulate emissions include: further restrictions on sulphur in fuel limits most likely in Polar regions; internal engine requirements by improving combustion; and, post-engine abatement in particular SCR and cyclones [28].

Regulation of greenhouse gas emissions, principally CO2, is one area that is subject to ongoing societal concern and coupled with recent international commitments is likely to result in a significant amount of change to design and operations. This may be achieved by: design measures that include machinery improvements, hydrodynamic improvements and renewables; operational improvements that could be supplemented by energy management systems; and, moving to alternative lower carbon fuels including biofuels [29]. Whilst significant change is forecast to reduce CO2 emissions there is also significant uncertainty. The first step should be accurately understanding current ship fuel use (which is the first step for the EU MRV) then setting fuel efficiency requirements for future ships' design. Many approaches could be combined to achieve those fuel efficiency requirements when placed alongside other design and operational requirements. Future design could also use a 'fit for not with' approach that could include: engines and auxiliaries that can operate on alternative fuels; instrumentation that would provide the right data into an energy management system; space for future waste heat recovery equipment; or even space forward for wind assisted propulsion equipment. By reducing CO2 emissions the design should also achieve the operational benefits of reduced fuel usage, namely: improved range; less reliance on refuelling tankers; and, greater availability.

In summary, beyond the forthcoming global sulphur in fuel restrictions and NOx ECAs there doesn't appear to be any identifiable future trend in regulation of these emissions; although in the medium term there could be additional Emission Control Areas. Regulation of black carbon and particulate matter is expected in the future but given its current status this would be in the medium term. Reduction of CO2 emissions appears to be the area of greatest impact in the short to long term although how this will be achieved is subject to much uncertainty.

6. CONCLUSIONS

By evaluating trends in environmental regulation it's possible to identify future environmental design requirements. These cover management of oil, black and grey water emissions, SOx and NOx emissions, black carbon and particulates and significant trends for reduction of CO2 emissions. The detail of the timing and the exact design requirements are difficult to determine. This may result in implementation difficulties when environmental requirements are being traded against other firm ship design requirements.

What is clear from this assessment is that the overall trend is one of tightening environmental regulations and history indicates that it is a matter of when not if, they will be implemented. This leaves ship designers and customers the difficult task of setting environmental design requirements that will be robust for the long service lives of naval ships whilst ensuring operations can be undertaken without constraint. As mentioned we can use the concept of ideality [8] for environment which is a zero emissions ship. In the short and probably long term this does not look achievable however it is a good starting point for design trade-offs particularly for liquid and solid emissions. One approach to achieving a robust design is to 'fit for but not with'. This could include additional space around abatement equipment for upgrades or provision for space and weight for air emissions abatement equipment in uptakes. This is particularly important for those aspects of the design subject to most technological change that would result in improved environmental performance.

It has been identified that reduction of CO2 emissions from ships will undergo a significant amount of change over the next 20-30 years. Requirements to mitigate CO2 emissions should be established early in the design process alongside a mechanism to value the in-service benefits of CO2 emissions reduction, such as reduced fuel usage and improved range, against any early lifecycle CAPEX.

Through the example of the Type 26 Global Combat Ship it has been demonstrated that the impact of future environmental regulations can be mitigated. This was achieved through the early adoption of environmental requirements for the design.

Lastly, it is acknowledged that this paper could not evaluate all aspects of environmental requirements and is focussed on those addressed by MARPOL. Further work could be undertaken to assess other areas of environmental requirements that include transfer of invasive species, noise pollution and the wider subject of ship design that is resilient to a changing climate.

7. REFERENCES

1. GARI, L., 'Arabic Treatises on Environmental Pollution up to the End of the Thirteenth Century', Environment and History, Vol. 8, No. 4, 2002

2. MICHAL MCMAHON, A, 'Small Matters: Benjamin Franklin, Philadelphia and the Progress of Cities', The Pennsylvania Magazine of History and Biography, Vol. 116, No. 2, 1992

3. Ellen MacArthur Foundation, 'Circular Economy Overview', www.ellenmacarthurfoundation.org/circular-economy/overview/concept, 2017

4. SAMIULLAH, Y., 'Biological effects of marine oil pollution', Oil and Petrochemical Pollution, Vol.2, Iss. 4, 1985

5. MARINOS, E. & GIOLANDA-ALEXIA, S., 'The History of MARPOL', Merchant Marine Academy, A.E.N. Macedonia, 2017

6. VAUGHN, A., 'Torrey Canyon disaster – the UK's worst ever oil spill 50 years on', The Guardian, 2017

7. MISHRA, U., 'Introduction to the Concept of Ideality in TRIZ', http://www.trizsite.tk, 2013

8. ARCTIC COUNCIL, '2020 Future Scenario for the Northwest Passage and the Canadian Arctic', Arctic Marine Shipping Assessment Report, 2009

9. HANSEN, C. et al, 'Arctic Shipping – Commercial Opportunities and Challenges', CBS Maritime, 2016

10. KOBOEVIC, Z., KOMADINA, P. & KURTELA, Z., 'Protection of the Seas from Pollution by Vessels' Sewage with Reference to Legal Regulations', Promet – Traffic & Transportation, Vol. 23, No. 5, 2011

11. THE GREEN BLUE, 'Sewage and Waste Water from Boats', Boating Factsheet 2, 2008

12. PRIOR, S., ASOC, 'Discharge of sewage and grey water from vessels in Antarctic Treaty waters', 36th Antarctic Treaty Consultative Meeting, 2013

13. CORBETT, J. & WINEBRAKE, J., 'The Impacts of Globalisation on International Maritime Transport Activity', Global Forum on Transport and Environment in a Globalising World, 2008

14. GEYER, R., JAMBECK, J. & LAVENDER LAW, K., 'Production, use, and fate of all plastics ever made', Science Advances, Vol. 3, No. 7, 2017

15. LAVENDER LAW, K., 'Plastics in the Marine Environment', Annual Review of Marine Science, Vol. 9, 2017

16. BBC REWIND, 'Blue Planet II: The clean-up', https://canvas-story.bbcrewind.co.uk/blue-planet-two-six-months-on/, 2018

17. GREENPEACE INTERNATIONAL, 'Greenpeace expedition finds plastic pollution and hazardous chemicals in remote Antarctic waters', https://www.greenpeace.org/international/press-release/16913/greenpeace-expedition-finds-plastic-pollution-and-hazardous-chemicals-in-remote-antarctic-waters/, 2018

18. United Nations Environment Programme, 'Single-use plastic has reached the world's deepest ocean trench', https://www.unenvironment.org/news-and-stories/story/single-use-plastic-has-reached-worlds-deepest-ocean-trench, 2018

19. EUROPEAN COMMISSION, MEMO: '72nd session of the Marine Environment Protection Committee (MEPC 70) at the International Maritime Organization (IMO)', 2018

20. United Nations Environment Programme, 'As the EU weighs a ban on single-use plastics', these European countries are moving ahead, https://www.unenvironment.org/news-and-stories/story/eu-weighs-ban-single-use-plastics-these-european-countries-are-moving-ahead, 2018

21. UK Government, 'UK Government rallies Commonwealth to unite on marine waste', https://www.gov.uk/government/news/uk-government-rallies-commonwealth-to-unite-on-marine-waste, 2018

22. VIANA M. et al, 'Impact of maritime transport emissions on coastal air quality in Europe', Atmospheric Environment, Vol. 90, 2014

23. STOCKER, T. et al (eds.), 'Climate Change 2013: The Physical Science Basis: Contribution of Working Group I to the Fifth Assessment Report of the Intergovernmental Panel on Climate Change', 2013

24. HARVEY, F., 'Shipping industry criticised for failure to reach carbon emission deal', The Guardian, 2016

25. RAHM, S., 'Schroders: The costly future of green shipping', 2015

26. MARSH REPORT, 'Emission Regulations: Concerns for the Marine Industry', 2018

27. DNV-GL MARITIME, DNV GL NOx Tier III Update – Choices and challenges for on-time compliance, 2017

28. DNV-GL MARITIME, 'Particulate Matter and Black Carbon: The global impact and what this means for shipping', 2016

29. DNV-GL MARITIME, 'Low Carbon Shipping Towards 2050', 2017

8. AUTHORS BIOGRAPHY

James Livingstone holds the current position of Type 26 Lead Environmental Engineer at BAE Systems Naval Ships. He is responsible for facilitating and delivering the environmental requirements for the Type 26 Global Combat Ship. Responsibilities cover Project subject matter expertise on environmental matters, delivery of training, design review and supplier deliverable review. Previously as an environmental consultant he has delivered support and technical advice to UK MOD project teams across all domains and to the wider defence estate.

9. FIGURES

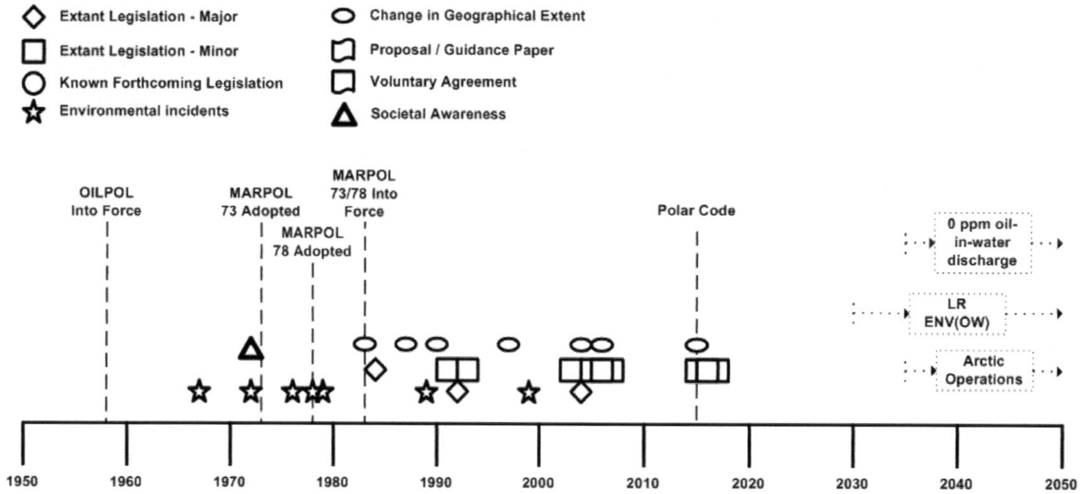

Figure 1: Pollution from Oil - History and Potential Future Trends

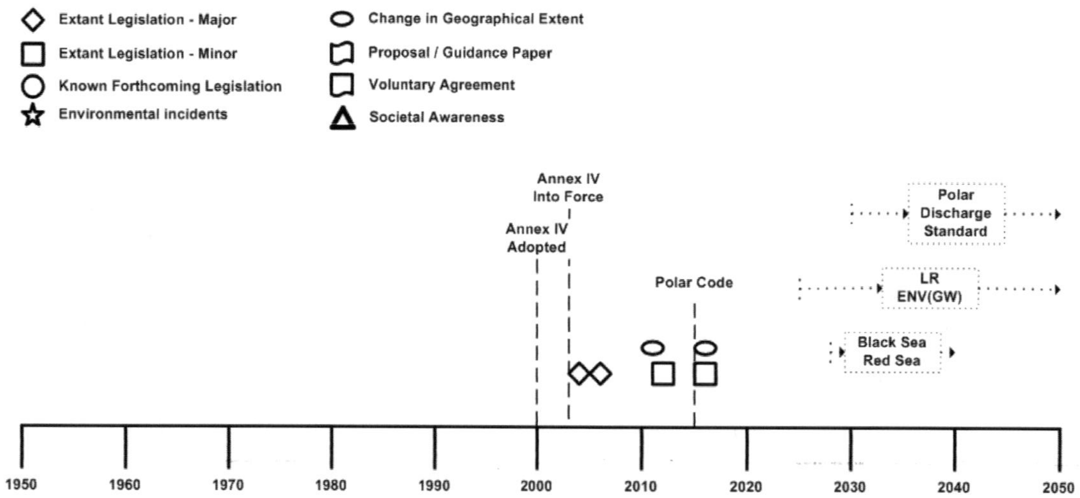

Figure 2: Pollution from Sewage and Grey Water - History and Potential Future Trends

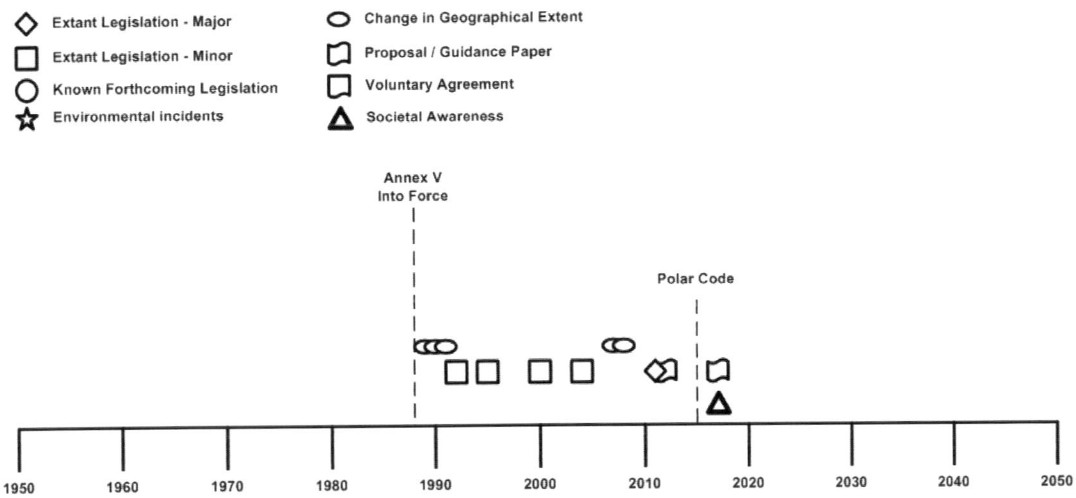

Figure 3: Pollution from Waste - History and Potential Future Trends

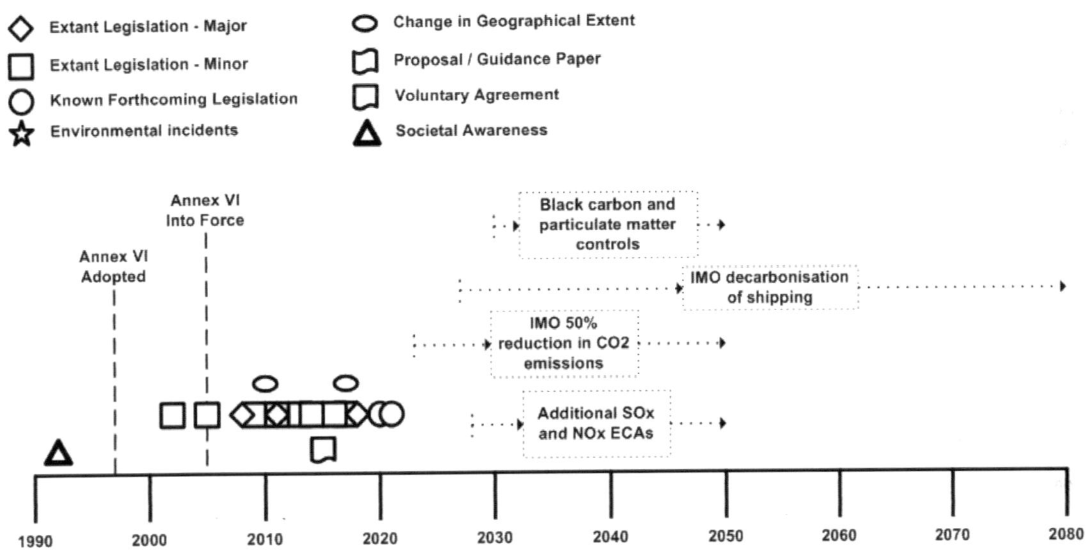

Figure 4: Pollution from Air Emissions - History and Potential Future Trends

AUTHORS' CONTACT DETAILS

Henry Parker
Ministry of Defence
Ben.Kadinopoulos830@mod.gov.uk

Oliver Short
Naval Design Partnering Team
Ollie.Short@babcockinternational.com

Rachel Pawling
University College London
r.pawling@ucl.ac.uk

Stuart Hunt
BAE Systems
stuart.hunt2@baesystems.com

K Vignesh Kumar
Indian Institute of Technology Delhi
kvkumar@am.iitd.ac.in

Alex Aitken
BMT
AAitken@bmtdsl.co.uk

Chris Broadbent
Ministry of Defence
CNBROADBENT@mail.dstl.gov.uk

Harry Schweidler
Babcock International Group
Harry.Schweidler@babcockinternational.com

David Andrews
University College London
d.andrews@ucl.ac.uk

Lewis Griffiths
BMT
LGriffiths@bmtdsl.co.uk

Jake Rigby
BMT
JRigby@bmtdsl.co.uk

James Schofield
Survivability Consulting Limited
jschofield@survivability.co.uk

David Manley
Ministry of Defence
David.Manley641@mod.gov.uk

Malcolm Courts
BAE Systems
malcolm.courts@baesystems.com

Pilar Morales
Babcock International Group
Pilar.Morales@babcockinternational.com

Cody Lyster
Babcock International Group
Cody.Lyster@babcockinternational.com

Ashley James
Babcock International Group
Ashley.james@babcockinternational.com

James Clarke
Babcock International Group
James.Clarke@babcockinternational.com

James Livingstone
BAE Systems
james.livingstone@baesystems.com

Mark Sansom
Babcock International Group
Mark.Sansom@babcockinternational.com

Veronica Williams
Babcock International Group
Veronica.Williams@babcockinternational.com

The Royal Institution of Naval Architects
8-9 Northumberland Street
London, WC2N 5DA
United Kingdom

ISBN 978-1-5108-8323-9

CURRAN ASSOCIATES INC.
proceedings.com

34th European Peptide Symposium and the 8th International Peptide Symposium 2016

Leipzig, Germany
4-9 September 2016